Inequality, Economic Growth, and Technological Change

Contributions to Economics

Giuseppe Gaburro (Ed.)
Ethics and Economics
1997. ISBN 3-7908-0986-1

Frank Hoster/Heinz Welsch/
Christoph Böhringer
CO_2 Abatement and Economic Structural Change in the European Internal Market
1997. ISBN 3-7908-1020-7

Christian M. Hafner
Nonlinear Time Series Analysis with Applications to Foreign Exchange Rate Volatility
1997. ISBN 3-7908-1041-X

Sardar M.N. Islam
Mathematical Economics of Multi-Level Optimisation
1998. ISBN 3-7908-1050-9

Sven-Morten Mentzel
Real Exchange Rate Movements
1998. ISBN 3-7908-1081-9

Lei Delsen/Eelke de Jong (Eds.)
The German and Dutch Economies
1998. ISBN 3-7908-1064-9

Mark Weder
Business Cycle Models with Indeterminacy
1998. ISBN 3-7908-1078-9

Tor Rødseth (Ed.)
Models for Multispecies Management
1998. ISBN 3-7908-1001-0

Michael Carlberg
Intertemporal Macroeconomics
1998. ISBN 3-7908-1096-7

Sabine Spangenberg
The Institutionalised Transformation of the East German Economy
1998. ISBN 3-7908-1103-3

Hagen Bobzin
Indivisibilities
1998. ISBN 3-7908-1123-8

Helmut Wagner (Ed.)
Current Issues in Monetary Economics
1998. ISBN 3-7908-1127-0

Peter Michaelis/Frank Stähler (Eds.)
Recent Policy Issues in Environmental and Resource Economics
1998. ISBN 3-7908-1137-8

Jessica de Wolff
The Political Economy of Fiscal Decisions
1998. ISBN 3-7908-1130-0

Georg Bol/Gholamreza Nakhaeizadeh/Karl-Heinz Vollmer (Eds.)
Risk Measurements, Econometrics and Neural Networks
1998. ISBN 3-7908-1152-1

Joachim Winter
Investment and Exit Decisions at the Plant Level
1998. ISBN 3-7908-1154-8

Bernd Meyer
Intertemporal Asset Pricing
1999. ISBN 3-7908-1159-9

Uwe Walz
Dynamics of Regional Integration
1999. ISBN 3-7908-1185-8

Michael Carlberg
European Monetary Union
1999. ISBN 3-7908-1191-2

Giovanni Galizzi/
Luciano Venturini (Eds.)
Vertical Relationships and Coordination in the Food System
1999. ISBN 3-7908-1192-0

Gustav A. Horn/
Wolfgang Scheremet/
Rudolf Zwiener
Wages and the Euro
1999. ISBN 3-7908-1199-8

Dirk Willer
The Development of Equity Capital Markets in Transition Economies
1999. ISBN 3-7908-1198-X

Karl Matthias Weber
Innovation Diffusion and Political Control of Energy Technologies
1999. ISBN 3-7908-1205-6

Heike Link et al.
The Costs of Road Infrastructure and Congestion in Europe
1999. ISBN 3-7908-1201-3

continued on page 188

Volker Grossmann

Inequality, Economic Growth, and Technological Change

New Aspects in an Old Debate

With 11 Figures and 5 Tables

Springer-Verlag
Berlin Heidelberg GmbH

Series Editors
Werner A. Müller
Martina Bihn

Author

Dr. Volker Grossmann
University of Zurich
Department of Economics
Rämistraße 62
CH-8001 Zurich
Switzerland
Email: volker.grossmann@wwi.unizh.ch

ISSN 1431-1933
ISBN 978-3-7908-1364-7

Cataloging-in-Publication Data applied for
Die Deutsche Bibliothek – CIP-Einheitsaufnahme
Grossmann, Volker: Inequality, economic growth, and technological change: new aspects in an old debate; with 5 tables / Volker Grossmann. – Heidelberg; New York: Physica-Verl., 2001
Zugl.: Regensburg, Univ., Diss.
ISBN 978-3-7908-1364-7 ISBN 978-3-642-57594-5 (eBook)
DOI 10.1007/978-3-642-57594-5

Softcover Design: Erich Kirchner, Heidelberg

SPIN 10790534 88/2202-5 4 3 2 1 0 – Printed on acid-free paper

To my parents

Eberhard and Gerda

Preface

Do more equal economies grow faster? Which role does the public sector play for the relationship between inequality and growth? What is the impact of technological changes and capital accumulation on the distribution of wealth, income, and wages? How do social comparisons among individuals affect redistributive taxation, the wage structure, and the growth rate of an economy? To answer these questions is the goal of this book.

I became interested in the issues of distribution and growth during my study of economics at the University of Bonn and as an exchange student at the UC Berkeley. Most notably, I was stimulated by my academic teachers George Akerlof, Paul Romer and Axel Weber.

This book was written during my stay at the department of economics at the University of Regensburg, in the years 1997-2000. It is my doctoral dissertation in economics (University of Regensburg; day of disputation was July 24th 2000). I am very grateful to my supervisors Wolfgang Buchholz and Josef Falkinger for their continuous support and very helpful suggestions on many drafts of parts of this book. I am also indebted to my colleagues Hartmut Egger and Martin Summer for illuminating discussions as well as to participants of many conferences and research seminars where parts of this book have been presented. Last but not least, I would like to thank my partner Ronke Osikominu for her moral support and my parents Gerda and Eberhard for more than I can say.

Volker Grossmann　　　　Regensburg, September 2000

Contents

Introduction

For many decades, it has been proclaimed that economic development and technological progress should be on top of the political agenda. However, there also have been worries that fast growth would lead to socially unacceptable levels of income/wealth inequality and even to increased poverty (i.e. to a decline not only of the relative position of a bottom share of the population but also of its absolute position). The European welfare states which are in place today may be regarded as a recognition of these concerns. However, large-scale redistribution measures have increasingly come under attack. One argument to cut welfare expenditure is that redistribution of income and wealth towards the poor distorts savings and investment decisions (since, for instance, a tax on human or physical capital income reduces the return to capital). It has also been suggested that an unequal society serves to provide individual incentives to work hard, to accumulate skills and to undertake risky projects, since the marginal benefit per unit of effort in these activities is high when the income ladder to climb is steep.[1] Moreover, under certain conditions, if the marginal propensity to save is rising in income, a more unequal distribution of income boosts savings and thus possibly investment and growth.[2] Because of these arguments, it has been concluded that redistribution has a negative impact on growth. Moreover, the views that on the one hand growth fosters inequality and on the other hand inequality yields higher effort and higher aggregate investment suggest a positive relationship between inequality and growth.

According to the two sides of the inequality-growth relationship the literature may be classified into two categories. First, the literature on the impact of capital accumulation and technological change on the distribution of income and wealth, and second, the literature on the impact of inequality on the growth rate of an economy.

Concerning the first category, one has to go back to both the classical economists (e.g. Adam Smith and David Ricardo) and Karl Marx who were mainly interested in the *functional income distribution*. Today, there is not even a consensus about whether or not the functional income distribution has changed in Western economies of the 20th century. Recently, due to the (particularly Anglo-American) experience of increased *personal income inequality* in the last, say, two decades, economists have mainly discussed

---

[1] See Welch (1999) for a recent discussion, approving this view.

[2] See Bourguignon (1981) for a welfare analysis of stationary egalitarian and unegalitarian income distributions in a neoclassical growth model in which convex saving functions are assumed.

the determinants of the personal (rather than the functional) income distribution.[3] One center of this debate is the so-called *skill-biased technological change* which is usually understood as a technology-related increase in the labor demand for better qualified workers relative to the demand for less-skilled workers. According to another notion, the accumulation of physical capital raises the productivity of high-skilled labor more than that of less-skilled labor. Both channels potentially lead to higher earnings inequality (and thus possibly higher income inequality) as the economy grows. Moreover, Kuznets (1955) developed his famous hypothesis that inequality first rises and eventually falls in the development process of an economy. Recently, this hypothesis has been addressed in the literature on *trickle-down effects* of growth.[4] Trickle-down growth is said to occur when inequality decreases in the process of wealth accumulation.
The second category encompasses the literature on the impact of personal income/wealth inequality on growth which should be positive according to the arguments above. However, recent empirical estimates indicate a fairly robust negative impact of income inequality on the growth rate of national income in cross-country studies.[5] Moreover, empirical studies do not find any conclusive relationship between redistribution and growth.[6] Whereas macroeconomic theory of the 1970s and 1980s has been dominated by representative agent models which, by design, could not contribute to understand the relationship between inequality and growth, the emergence of both the empirical literature on this subject and endogenous growth theory has stimulated an enormous amount of theoretical research in the 1990s.

This book focuses on the question whether there is a positive or negative relationship between the income/wealth distribution and an economy's rate of growth. In line with the two categories outlined above, the attempts to answer this question go in the following directions. First, the impact of inequality and redistribution on growth is examined when growth is driven by human or physical capital investments. Second, the impact of

---

[3] Of course, the investigation of the personal income distribution started much earlier and at least dates back to the end of the 19th century when Wilfredo Pareto discovered that the distribution of income seems to follow a regular pattern, which today is known as 'Pareto's law'. For a discussion of this and other attempts to find empirical regularities and theoretical explanations of the income distribution, see e.g. the books of Pen (1971) and Kakwani (1980).
[4] E.g. Aghion and Bolton (1997), Piketty (1997).
[5] E.g. Alesina and Rodrik (1994), Persson and Tabellini (1994).
[6] E.g. Easterly and Rebelo (1993), Perotti (1996).

technological progress and accumulation on the distribution of earnings and income is investigated.

Moreover, the role of *equity concerns* and *social comparisons*, respectively, for macroeconomic variables is examined. As already indicated above, income inequality is often regarded as socially unjust. But equity concerns undoubtedly are present on the individual level as well. Probably not many people would deny that individuals care about their relative standing in some reference group and hold rather well-defined perceptions about fair distributions of material (and non-material) goods. The analysis in this book takes such social comparisons and equity concerns of individuals seriously. If status and fairness concerns are important determinants of human motivation, it is natural to examine their impact on the macroeconomic performance of an economy. For instance, if, as often suggested, a trade-off between income equality and growth is present, a society's solution to this trade-off may be largely affected by those equity concerns.

The book attempts to serve two purposes. First, a broad and balanced picture about the relationship between inequality, growth and technological progress should be provided. Even though one may reject this goal as too ambitious due to the complexity of the issues involved, this may nevertheless be a worthwhile attempt precisely *because* of the various interactions between inequality, growth and technological innovations. That is, it may actually be necessary to take a broad view in order to come to a differentiated picture regarding these interactions as well as to unbiased conclusions with respect to economic policy. Second, the book not only tries to contribute to the discussion on a general level but also to introduce specific aspects which emerge from this discussion. Roughly spoken, this amounts to new results about the impact of equity concerns in the context of growth, to a more differentiated picture about the individual demand for redistribution policies and the size of the public sector when public expenditure is financed by growth-depressing income taxes, and to a novel explanation for the decline in relative earning opportunities of less-skilled workers.

These purposes lead to the following plan of the book. Chapter 1 provides an update of the fast growing theoretical and empirical literature on the relationship between inequality, growth and technical progress. The survey especially focuses on the question whether or not a more equal income distribution and redistributive activities depress growth, and whether or not technological progress and capital accumulation lead to a more unequal income and earnings distribution. In order to discriminate between the large number of theoretical results, empirical results are extensively reviewed as well. Moreover, some new cross-section data addressing the famous Kuznets hypothesis is provided. The discussion in chapter 1 then motivates the analysis in chapters 2-4.

Chapters 2 and 3 look at the redistributive process via the public sector in a democracy when government activities are financed through factor income taxation. We consider models, in which taxes levied on income from accumulated production factors depress investment-driven growth. Thus, in these models, there is a trade-off between equality and growth which has to be resolved by individuals in shaping their demands for public policy.

Chapter 2 analyzes the relationship between status concerns, inequality and growth. A model is presented in which individuals not only care about their *absolute* level of consumption (like in textbook models), but also about their relative standing in society, i.e. their *relative* level of consumption. It is first argued that there is convincing evidence for the notion that social interaction creates individual incentives to strive for social status. Having established this, the impact of relative consumption preferences on the rate of growth is considered. Two channels are distinguished. First, the impact of preferences for the relative level of individual consumption on saving incentives (for a given redistribution policy), and second, their impact on voting behavior with respect to a redistributive tax-transfer scheme (where redistribution depresses growth). Moreover, it is examined how the income distribution affects growth in the presence of status concerns.

Chapter 3 provides an explanation for the empirical finding that unequal economies do not have higher tax rates than more equal ones (i.e. that the poor do not seem to expropriate the rich through the tax system). Redistribution and taxation have been suggested to play an important role for the question how inequality adversely affects an economy's rate of growth. (The negative relationship between inequality and growth and possible explanations are discussed in chapter 1). In the model of chapter 3, as in the one of chapter 2, income taxes are levied in a growth-reducing manner. But now the tax revenue is used for the public provision of (public or private) goods and services rather than for direct income transfers. Due to the assumption that all individuals consume the same amount of these goods and services although being charged (i.e. taxed) differently, this kind of government action has a redistributive character. It is examined under which circumstances richer individuals may nevertheless prefer larger sizes of the public sector than poorer ones. It is argued that, in contrast to the conventional wisdom, in an equilibrium in which the tax rates are determined by majority voting inequality is generally not positively related to the level of taxation. As taxation adversely affects growth, one can thus conclude that inequality is not negatively related to growth through the political process. First, a static model is analyzed. Second, the additional effects emerging from savings decisions are worked out in a growth model (similar to the one in chapter 2) when richer individuals have higher propensities to save. It is shown that different saving propensities among individuals play an important role to understand how individuals adjust their private consumption levels to taxation, i.e. the propensity to save is shown to be an important determinant for the individual trade-off between private and public consumption when demand for public policy is shaped.

Chapter 4 deals with the impact of technological change on income (i.e. earnings) inequality. Whereas in the models of chapter 2 and 3 productivity growth (i.e. technological change) is due to socially non-decreasing returns to (human and physical) capital (i.e. growth is investment-driven),[7] in chapter 4 technological change depends on incentives for research firms to innovate (i.e. to find blue-prints for productivity-enhancing ways in the consumption goods sector). These research activities should be understood as to represent any activity undertaken to boost productivity and/or the revenue of firms. Thus, one may also think about consulting, marketing, the reorganization of the hierarchical structure of a firm, the reorganization of work in teams, supervision of workers, etc. In fact, the empirical literature on skill-biased technological change suggests not only that such *non-production activities* have quantitatively become much more important, but also that they are associated with the decline in relative demand for low-skilled labor and thus with an increase in earnings inequality. With respect to the last point, to my knowledge, the entire literature on this topic lacks a formal model of those non-production activities. (See the survey in chapter 1). Usually, skill-biased technological change is formalized by a shift in the production function that makes better qualified workers relatively more productive. As shown in chapter 4, analyzing a model which distinguishes between production and non-production activities (in the above sense), and relating it to the empirical literature on the link between technological change and labor demand for differently qualified workers, provides an alternative view for a rising relative demand for skills. Moreover, whereas chapter 2 starts out with the assumption that individuals are motivated by *status concerns*, in chapter 4 it is assumed that the *wage demands* of workers are affected by *fairness concerns*. As has been shown by Akerlof and Yellen (1990), fairness considerations *between* high-skilled and low-skilled workers give rise to a trade-off between wage equality and employment of low-skilled workers. That is, fairness concerns lead to a *compression of wages* of high-skilled relative to low-skilled workers in the sense that the wage differential between the two types of workers is lower than the productivity differential. In addition, chapter 4 also considers the distributional and employment consequences of social comparisons *within*

---

[7] Positive externalities of capital accumulation may arise from knowledge spill-over or learning-by-doing effects, respectively, as in Arrow (1962). Romer (1986) takes up this idea in his seminal paper on endogenous growth.

the groups of equally skilled workers who conduct different activities. It will be shown that labor market effects of social comparisons depend on the interaction between relative wage stickiness arising from fairness considerations between the groups of high-skilled and low-skilled workers and absolute wage stickiness arising from fairness considerations among the low-skilled. Thus, by taking into account social comparisons, a differentiated picture of the macroeconomic effects of technological change is worked out.

Chapter 5 summarizes the results of the preceding chapters, derives policy implications, outlines open questions and suggests further research possibilities.

Chapter 1. Where do we stand? A survey

In this chapter, the literature on the link between income/wealth distribution, economic growth and technological progress is reviewed, emphasizing open questions and unexplored issues which are analyzed in the subsequent chapters.

The survey is organized and motivated as follows. Section 1.1 briefly reviews the *classical and Marxian perspective* of the question how capital accumulation and technological progress can affect the *functional income distribution*. The classical view is mainly represented by David Ricardo[1] although some of his ideas build on Adam Smith.[2] In comparison with this view, the pioneering work of Karl Marx[3] is discussed. This is followed by a brief discussion about the distributive aspects of both *Keynesian and neoclassical growth theory* (section 1.2). Section 1.3 revisits the *Kuznets hypothesis* and confronts it with some new cross-section data. The goal of these three sections is to outline relevant questions in the inequality and growth debate from the perspective of the history of economic thought, and to demonstrate that most of the issues which are discussed nowadays have long been addressed by economists. The main focus of the survey, however, is the more recent literature about inequality and growth, reviewed in sections 1.4 and 1.5, respectively.

Section 1.4 deals with the *impact of inequality and redistribution on the rate of growth*. Some of the outlined models not only consider how the rate of growth depends on some initial level of inequality but also how inequality changes with the accumulation of wealth. The central questions in this section are the following. (i) What are the relevant channels that can explain the empirically observed negative relationship between income inequality and an economy's rate of growth? (ii) How can one empirically discriminate between the various theoretical explanations? (iii) How do redistribution policies affect growth? (iv) Do more unequal economies have more or less redistribution? The discussion of these questions is supposed to motivate the analysis of the new aspects addressed in chapters 2 and 3.

---

[1] Ricardo (1821 [reprinted 1965], *The principles of political economy and taxation*, 3rd edition. 1st edition published 1817).

[2] Smith (1776 [reprinted 1937], *An inquiry into the nature and causes of the wealth of nations*).

[3] Marx (1890 [reprinted 1969], *Das Kapital I*, 4th edition. 1st edition published 1867).

Section 1.5 primarily deals with the *impact of technological progress on earnings inequality.*[4] That is, various technology-related explanations as well as empirical evidence for both the increasing wage dispersion *across* observably different groups of workers (grouped by e.g. education or occupation) and the increasing wage dispersion *within* observably similar groups in the last, say, two decades are discussed. But also recent trends of both the functional income distribution and the personal income distribution are outlined. To focus on earnings inequality in section 1.5 follows the recent debate about the impact of growth on inequality (which is different to the focus of the older literature, as will be seen from the discussion in sections 1.1-1.3). As roughly two thirds of national income is paid in wages, it is reasonable to hypothesize that changes in the wage distribution have a crucial impact on the personal income distribution.[5] The evaluation of the literature on the wage dispersion across groups is supposed to motivate an alternative view presented in chapter 4.
The literature about inequality, growth and technical progress has been growing very fast in the last decade. This survey should give a flavor for the complex interactions between income distribution, capital accumulation and technological innovations. And, it should underline the importance of a broad understanding of these topics for both macroeconomic theory and economic policy.

1.1 Growth and the functional income distribution: the classical and Marxian perspective

Adam Smith, David Ricardo and Karl Marx argued that capital accumulation and technological change are main determinants of the functional income distribution.[6] In order to avoid going into too much detail,

---

[4] Recently, Aghion, Caroli and García-Peñalosa (1999) have surveyed the literature on the relationship between inequality and growth along similar lines as in sections 1.4 and 1.5.

[5] It has to be noted that the theoretical literature on the distribution of personal income which is due to capital income/wealth inequality is not captured in section 1.5. The literature on this issue is reviewed in section 1.4 when the so-called *trickle-down growth* models are discussed. (Trickle-down growth is said to occur when income/wealth inequality falls in the process of development). These models aim to analyze the simultaneous evolution of growth and inequality, given an initially unequal wealth distribution.

[6] The interpretations in this section mainly draw on Kaldor (1956), Adelman (1961) and Mandel (1987). See also Schumpeter (1965).

the discussion of the classical and Marxian perspective is restricted to their pioneering work, and refers only to the most essential features of the relationship between inequality and growth their theories imply. Interestingly, and in line with an important goal of this book, it is also argued that both Ricardo and Marx acknowledged the role of *social comparisons* for macroeconomic variables.[7]

1.1.1 The classical view

Both Smith (1776 [reprinted 1937]) and Ricardo (1821 [reprinted 1965]) distinguished between the three production factors labor, capital and land. Ricardo explicitly aimed to explain changes in the functional income distribution, i.e. he viewed the labor share, the capital share and the land rent share of total income as endogenous variables.[8]

> [T]he produce [...] is divided among three classes of the community, namely, the proprietor of land, the owners of the stock or capital necessary for its cultivation, and the labourers by whose industry it is cultivated. But in different stages of the society, the proportions of the whole produce of the earth which will be allotted to each of these classes, under the names of rent, profits, and wages, will be essentially different; depending mainly on the actual fertility of the soil, on the accumulation of capital and population, and on the skill, ingenuity, and the instruments employed in agriculture. [Ricardo (1965, preface)].

Smith, as Ricardo, saw capital accumulation as a necessary condition for an increase in labor productivity, mainly through its effect on the ''division of labor''.

> As the accumulation of stock must, in the nature of things, be previous to the division of labour, so labour can be more and more subdivided in proportion only as stock is previously more and more accumulated. [Smith (1937, p. 260)].

---

[7] However, it should be noted that, rather than on equity concerns at an individual level, Marx especially focused on the role of the *social organization* in a capitalistic system. According to Marx, capitalism is characterized by an inevitable ''elementary class struggle and elementary unionization of the working class'' as well as a ''growing social polarization'' [Mandel (1987, p. 380)].

[8] It is interesting to note that the determinants of growth and labor productivity which Ricardo pointed out are exactly the ones which play a role in the so-called 'new growth theory', i.e. physical and human capital accumulation, knowledge and innovations. Seminal papers on the new growth theory are Romer (1986, 1990), Lucas (1988), Grossman and Helpman (1991), Aghion and Howitt (1992).

However, both writers argued that the long run real wage rate would not depend on the productivity of labor. Rather, there would be a "natural price of labour which is necessary to enable the labourers [...] to subsist", depending on "the price of the food, necessaries, and conveniences required for the support of the labourer and his family" [Ricardo (1965, p. 52)]. Smith viewed this wage level as "subsistence wage" in the physiologic sense, i.e. as one which ensures survival.[9] In contrast, Ricardo pointed out that "the natural price of labour [...] would vary at different times in the same country, and would materially differ in different countries". That is, rather than being fixed, the real wage "essentially depends on the habits and customs of people" (pp. 54-55). Thus, over time, real wages would change due to changes in workers' subjective view about the subsistence level of wages.

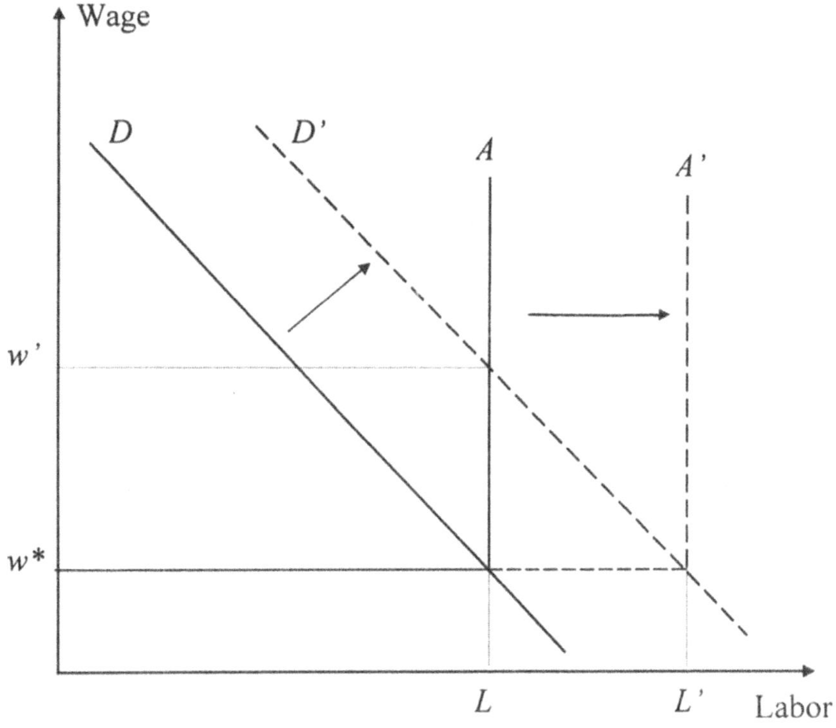

Figure 1.1: The impact of capital accumulation on the labor market: The Ricardian view.

[9] "Every species of animals naturally multiplies in proportion to the means of their subsistence and no species can ever multiply beyond it" [Smith (1937, p. 79)].

Labor demand would crucially depend on the capital stock, i.e. on "food, clothing, tools, raw materials, machinery, etc." which would be "necessary to give effect to labour" (p. 53).[10] As capital accumulates, real wages would rise only in the short-run. In the longer-run, there would be a "great addition [...] to the population" (e.g. due to better nutrition of children) such that "the market price of labour will again sink to its [...] natural price" (p. 54). This process is depicted in Fig. 1.1. According to Ricardo, labor supply is perfectly elastic. Labor demand D shifts rightwards to D' after an increase in the capital stock which induces a real wage increase from the subsistence wage w^* to w'. This causes the (working age) population to increase from L to L' yielding a new labor supply curve, i.e. a shift from A to A'. Thus, the new full employment equilibrium wage is, again, given by w^*.[11]

Ricardo believed in the law of diminishing returns (unlike Smith, who did not make any reference to it), i.e. that labor productivity would decline with an employment expansion unless there would be technological progress.[12] Thus, the main implication of Ricardo's theory for the functional income distribution is the following. Unlike in neoclassical (i.e. marginalist) theory, the marginal product of labor (which is smaller than the average product) is not equal to the wage rate. Rather, the marginal product of labor times total employment is equal to the sum of the wage-bill and profits. The remaining residual of the "whole produce" (i.e. the difference between the average and marginal product of labor times total employment) is equal to the rent appropriated by owners of land. Moreover, with constant real wages and a declining average productivity of labor, the share of labor increases with capital accumulation.[13]

---

[10] According to Smith (1937, p. 69), labor demand also depends on total output: "The demand for those who live by wages, it is evident, cannot increase of the funds which are destined for the payment of wages. These funds are of two kinds: first, the revenue which is over and above what is necessary for the maintenance; and, secondly, the stock which is over and above what is necessary for the employment of their masters".

[11] However, as De Vivo (1987) points out, Ricardo did not see that allowing the subsistence wage level to depend "on the habits and customs of people" means that an increase in actual wages ("market price of labour") due to capital accumulation affects subsistence wages ("natural price of labour") as well.

[12] " [...] the employment of an additional quantity labour with a proportionally less return" [Ricardo (1965, p.37)]. See Adelman (1961, p. 55), who argues that, in addition, "Ricardo believed that technological progress is also subject to diminishing returns" implying that "the long-run trend of profits will be downward".

[13] Formally, denote the fixed real wage as w^*, the number of workers (who are fully employed in equilibrium) as L and output (i.e. real national income) as Y.

According to the classical view, capital accumulation is a positive function of the return to capital. Since the "natural tendency of profits is to fall", eventually "the very low rate of profit will have arrested all accumulation".[14] This stationary state of the economy would be characterized by a constant population since "no capital can yield any profit whatever, and no additional labour can be demanded, and consequently population will have reached its highest point". As Adelman (1961, p. 57) points out, this does not imply, however, that all economies are reaching the same stationary state in Ricardo's model since the "level of subsistence required by the workers" (which, affects 'profits' and thus the development path of an economy) "is, in part, a socio-cultural phenomenon".
Rather than discussing cross-country differences in stationary states further,[15] one should note that Ricardo was probably the first economist who acknowledged a crucial impact of social factors on economic variables. Unfortunately, he was not explicit about the underlying mechanism how the wage demands by workers are shaped. However, according to the quoted passage above, the "habits and customs of people" certainly depend on the stage of development, i.e. on the average level of wealth or consumption in an economy, respectively. As suggested by Duesenberry (1949), if the subjective view about the basic needs of an individual (i.e. the individual consumption or wage level, respectively, which is necessary for subsistence) depends on the average consumption level in the reference group of that individual, this is equivalent to individual concerns about the *relative standing* in that group. That is, if others (e.g. capitalists) get richer, workers' wage demands (i.e. the subjective view about the subsistence wage) increase due to social comparisons (e.g. workers may socially compare themselves with capitalists or other workers, respectively). As outlined in the introduction, our analysis in chapter 4 deals with macroeconomic effects of social comparisons with respect to wages in the context of growth.

1.1.2 The theory of Karl Marx

In contrast to Smith and Ricardo, Marx (1890 [reprinted 1969]) did not distinguish between the return to capital and land (i.e. between 'profits' and

---

According to Ricardo (1965), both L and Y rise with an increase in the capital stock and average labor productivity Y / L declines. Thus, the labor share in national income $w^* L / Y$ increases with capital accumulation.

[14] See Ricardo (1965, p. 71) for this and the following quotation.

[15] See Adelman (1961, pp. 57-59) for further discussion of stationary states in Ricardo's theory.

'rent'). The value-added in production which is not paid in wages is called the 'surplus value'. According to Marx, capitalism is inevitably characterized by a struggle between the ruling class (all those who live off surplus value) and the producing class (i.e. workers).
The theory of Marx differs to the classical perspective in several other respects as well. Whereas capital accumulation in the classical theory is determined by its rate of return only, in the theory of Marx, capital accumulation is inherent in a capitalistic society. This is due to the following two reasons. First, Marx argued that for capitalists capital accumulation would be an end itself rather than a mean to raise profits. According to his view, it is inherent in a capitalistic economy that capitalists accumulate capital in order to gain social esteem.

> Nur insoweit der Kapitalist personifiziertes Kapital ist, hat er einen historischen Wert [...]. Als Fanatiker der Verwertung des Werts zwingt er rücksichtslos die Menschheit zur Produktion um der Produktion willen, daher zu einer Entwicklung der gesellschaftlichen Produktivkräfte und zur Schöpfung von gesellschaftlichen Produktionsbedingungen, welche allein die reale Basis einer höheren Gesellschaftsform bilden können, deren Grundprinzip die volle und freie Entwicklung jedes Individuums ist. Nur als Personifikation des Kapitals ist der Kapitalist respektabel. Als solche teilt er mit dem Schatzbildner den absoluten Bereicherungstrieb. [Marx (1969, p. 540)].

In this sense, *status-seeking* by capitalists has an impact on growth via its impact on capital accumulation and thus, as will be seen below, on both the income distribution and unemployment. Second, Marx presumed economies of large scale production, implying that competition among capitalists would foster accumulation and would lead to large concentration processes.[16]

> Die größeren Kapitale schlagen die kleineren. Man erinnert sich ferner, daß mit der Entwicklung der kapitalistischen Produktionsweise der Minimalumfang des individuellen Kapitals wächst, das erheischt ist, um ein Geschäft unter normalen Bedingungen zu betreiben. Die kleineren Kapitale drängen sich daher in Produktionssphären, deren sich die große Industrie nur noch sporadisch oder unvollkommen bemächtigt hat. Die Konkurrenz rast hier im direkten Verhältnis zur Anzahl und im umgekehrten Verhältnis zur Größe der rivalisierenden Kapitale. Sie endet stets mit dem Untergang vieler kleiner Kapitalisten und Übergang ihrer Kapitale in die Hand des Siegers. [Marx (1969, p. 576)].

---

[16] See also Steindl (1952) for an attempt of a modern foundation.

The latter motive for capital accumulation is usually emphasized in reviews of the work of Marx, whereas the former is frequently neglected.

Like Smith and Ricardo, Marx assumed that workers would receive subsistence wages. As Mandel (1987) points out, Marx did not see subsistence wages to be tied to a physiological minimum, contrary to the interpretation of many other writers, since the inevitable 'class struggle' may well be sometimes victorious for the working class. However, unlike classical writers, Marx left no doubt about strong adverse effects of capital accumulation on both the income share of labor and the level of employment. As Kaldor (1956, p. 88) summarizes: "On the Marxian model the share of wages in output must necessarily fall with every increase in output per head". Moreover, whereas the classical view implies full employment (see Fig. 1.1), Marx presumed the presence of an unemployed 'reserve army'. As Adelman (1961, p. 70) points out, according to Marx, "any piece of capital equipment actually in operation requires fixed amounts of labor to work with". Thus, as the supply of labor is inelastic if the wage is not lower than a subsistence level (like in Fig. 1.1), unemployment is due to an inelastic labor demand which depends on the total stock of capital. As a result, the labor market does not clear and the wage is equal to a subsistence level. Does that imply that unemployment decreases over time when capital accumulates? This ceteris paribus implication of a fixed capital-labor ratio is strongly denied by Marx, because of two countering effects. First, (exogenous) population growth and second, technological change. Concerning the latter, Marx distinguished between capital accumulation (and concentration) without and with changes in the 'organic composition of capital', i.e. without and with technological progress. The 'organic composition of capital' is defined as the ratio of 'constant capital' (i.e. depletion of raw materials, depreciation of machinery and buildings) to 'variable capital' (i.e. the total wage-bill). According to Marx, this ratio rises with capital accumulation due to technological change and thus labor demand per unit of capital decreases over time. The net effect of capital accumulation on labor demand would be negative.

> Ein stets größerer Teil des Kapitals wird in Produktionsmittel umgesetzt, ein stets kleinerer Teil in Arbeitskraft. Mit dem Umfang, der Konzentration und der technischen Wirksamkeit der Produktionsmittel vermindert sich progressiv der Grad, worin sie Beschäftigungsmittel der Arbeiter sind. [...] Und während die Zentralisation so die Wirkungen der Akkumulation steigert und beschleunigt, erweitert und beschleunigt sie gleichzeitig die Umwälzungen in der technischen Zusammensetzung des Kapitals, die [...] die relative Nachfrage nach Arbeit vermindern. [...] Die im Verlauf der normalen Akkumulation gebildeten Zusatzkapitale [...] dienen vorzugsweise als Vehikel

> zur Exploitation neuer Erfindungen und Entdeckungen, überhaupt industrieller Vervollkommnungen. [...] Die hieraus notwendig folgende absolute Abnahme der Nachfrage nach Arbeit wird selbstredend um so größer, je mehr die diesen Erneuerungsprozeß durchmachenden Kapitale bereits zu Massen angehäuft sind vermöge der zentralisierenden Bewegung. [...] der Mechanismus der kapitalistischen Produktion sorgt dafür, daß der absolute Zuwachs von Kapital von keiner entsprechenden Steigerung der allgemeinen Arbeitsnachfrage begleitet ist. [Marx (1969, pp. 577, 579, 589-590)].

Moreover, as labor productivity rises but wages remain fixed (Marx called the gap between labor productivity and the wage rate the 'rate of exploitation'), the 'surplus value' increases. As Adelman (1961, p. 78) points out, according to Marx's theory, the working class as a whole loses because of a rise in unemployment as well. Finally, in Marx's system the overall well-being of workers is dramatically worsened while the economy grows because of deteriorating working and living conditions.

> Es folgt daher, daß im Maße wie Kapital akkumuliert, die Lage des Arbeiters, welches immer seine Zahlung, sich verschlechtert. [...] Es bedingt eine der Akkumulation von Kapital entsprechende Akkumulation von Elend. Die Akkumulation von Reichtum auf dem einen Pol ist also zugleich Akkumulation von Elend, Arbeitsqual, Sklaverei, Unwissenheit, Brutalisierung und moralischer Degradation auf dem Gegenpol, d.h. auf Seite der Klasse, die ihr eignes Produkt als Kapital produziert. [...] Je rascher das Kapital in einer in einer industriellen oder kommerziellen Stadt akkumuliert, um so rascher der Zustrom des exploitablen Menschenmaterials, um so elender die improvisierten Wohnlichkeiten der Arbeiter. [Marx (1969, pp. 595, 611)].

It should be noted that Marx acknowledged that wages could rise in the short-run as capital accumulates, even though they would eventually return to a subsistence level, as suggested by Smith and Ricardo.

> Da das Kapital jährlich einen Mehrwert produziert, [...] können die Akkumulationsbedürfnisse des Kapitals das Wachstum der Arbeitskraft oder der Arbeiteranzahl, die Nachfrage nach Arbeitern ihre Zufuhr beflügeln, und daher die Arbeitslöhne steigen. [Marx (1969, p. 563)].

However, the reason for an eventual fall in wages is very different to the classical perspective. Whereas Smith and Ricardo argued that population growth would depend on the level of wages, Marx argued that profits would decline with rising wages such that the accumulation process would come to a halt. In turn, this would cause wages to fall and the accumulation process to start again.

> [...] die Akkumulation erschlafft infolge des steigenden Arbeitspreises, weil der Stachel des Gewinns abstumpft. Die Akkumulation nimmt ab. Aber mit ihrer Abnahme verschwindet die Ursache ihrer Abnahme, nämlich die Disproportion zwischen Kapital und exploitabler Arbeitskraft. Der Arbeitspreis sinkt also wieder zu einem den Verwertungsbedürfnissen des Kapitals entsprechenden Niveau. [...] Der Mechanismus des kapitalistischen Produktionsprozesses beseitigt also selbst die Hindernisse, die er vorübergehend schafft. [Marx (1969, p. 569)].

For a general evaluation of Marx's work, one may follow Adelman (1961, p. 93) in concluding that

> his system lends itself quite easily to an investigation of the relationship between the character of technological change, the distribution of income through time, capital accumulation, and economic growth. [...] The inherently dynamic nature of Marx's models provides an excellent example of the power and importance of dynamic analysis.

Moreover, his notion of an increasing substitution of capital for labor, i.e. a rise in the 'organic composition of capital', is actually pretty similar to what today is discussed under the label of labor-saving technological progress, at least as far as low-skilled workers (in the manufacturing sector) are concerned. As will be discussed in section 1.5 and chapter 4 of this book, there is not much doubt that technological change has recently caused a decline in earning opportunities of low-skilled workers in particular. In most developed countries, the low-skilled experienced dramatic increases in unemployment, and, at least in the U.S., also a reduction in real wages. Consistent with theory of Marx, there is also evidence for a recent fall in the wage-bill share in national income. Critics of Marx have frequently claimed that his theory would have proven wrong due to a non-decreasing wage-bill share in capitalistic economies over time. However, as Mandel (1987, p. 379) points out:

> Living in the age of semi-automation and 'robotism', it is hard to understand that challenge [of Marx's theory]. The conceptual confusion on which this challenge is mostly based is [...] a confusion between wages in general and variable capital, which is only the wage bill of productive labour. A more correct index would be the part of the labour costs in total production costs per manufacturing (and mining) sector. It is hard to deny that this proportion shows a downward secular trend.

Finally, it is notable with respect to the purposes of this book, that Marx pointed to growth-effects of status-seeking. A modern perspective of this issue is depicted in chapter 2 of this book.

1.2 Distributional aspects of Keynesian and neoclassical growth theory

As pointed out above, both the classical and Marxian theory saw long-run wages to be tied to a subsistence level. In contrast, according neoclassical distribution theory,[17] if both factor and goods markets are perfectly competitive, the return to any production factor is equal to its marginal product. The first clear statements of this result can be found in Wicksteed (1894) and Clark (1899).[18] Wicksteed also showed that under constant returns to scale in production, the sum of factor payments is equal to total output (also called the 'adding up-theorem'). A famous example representing such a technology is the production function of Cobb and Douglas (1928), which implies that, if all factors are paid their marginal product, the income share of each factor is constant. That is, according to this technology, the functional income distribution is entirely determined by the parameters of this production function (which the authors have extensively estimated in their later work).
The demand-oriented (i.e. 'Keynesian') distribution theory is especially identified with the work of Nicholas Kaldor (1956, and later papers).[19] It is called 'Keynesian' since it rests on the assumption that investment decisions of entrepreneurs are determined by exogenous factors, rather than by savings and profits.[20] (According to Keynes (1936), these exogenous factors include

---

[17] This brief reference to the neoclassical distribution theory draws on Külp (1994) and Stigler (1941).

[18] Also Marshall (1895, p. 581) stated that an entrepreneur "estimates as best he can how much net addition to the value of his total product will be caused by a certain extra use of any one agent [...] and he endeavors to employ each to that margin at which its net product would no longer exceed the price he would have to pay for it". However he was cautious in saying that wages are equal to the marginal product of labor as he saw substantial difficulties of isolating the "net addition" of a factor. See e.g. Stigler (1941) for further discussion.

[19] See Külp (1994) for a discussion.

[20] As Kaldor (1956, p. 94) notes: "Keynes, as far as I know, was never interested in the problem of distribution as such. One may nevertheless christen a particular theory as 'Keynesian' if it can be shown to be an application of the specifically Keynesian apparatus of thought and if evidence can be adduced that at some stage in the development of his ideas, Keynes came near to formulating such a theory".

expectations and the technology). Under the assumption that the propensity to save is higher for entrepreneurs than for workers, and given that the propensity to invest is equal to the *total* propensity to save, Kaldor's theory implies that the income share of profits is an increasing function of the propensity to invest, and a decreasing function of the propensity to save of the group of entrepreneurs and workers, respectively. (For the wage-bill share the opposite results hold). That is, the functional income distribution is entirely determined by investment and saving behavior.
An exogenous propensity to invest is also assumed in the so-called Keynsian growth theory of Harrod (1939) and Domar (1946).[21] According to this theory, both the capital to output ratio and the labor to output ratio are fixed, implying that capital and labor are always used in a fixed proportion. In this scenario, a steady state requires that the 'warranted rate of growth', i.e. the propensity to invest divided by the exogenous capital to output ratio, equals the 'natural rate of growth', i.e. the exogenous population growth rate. As both rates are exogenous, there is no reason why this steady state condition should be fulfilled. However, if this is not the case in each period, the system diverges, either capital or labor is not fully employed and one factor price converges to zero.
This theoretically unappealing feature has led Robert Solow to the following critique and powerful modification of the Harrod-Domar model:

> A 'crucial' assumption is one on which the conclusions do depend sensitively, and it is important that crucial assumptions be reasonable realistic. When the results of a theory seem to flow specifically from a special crucial assumption, then if the assumption is dubious, the results are suspect. [...] But [the] fundamental opposition of warranted and natural rates turns in the end to flow from the crucial assumption that production takes place under conditions of fixed proportions. There is no possibility of substituting labor for capital in production. If this assumption is abandoned, the knife-edge notion of unstable balance seems to go with it. [Solow (1956, p. 65].

That is, by replacing the assumption of "fixed proportions" of capital and labor by a neoclassical production function with constant returns to scale which exhibits substitution possibilities between capital and labor, and letting each factor price be equal to the marginal productivity of that factor, Solow (1956) shows that the system converges to a stable steady state with full employment of both factors [see also Swan (1956)]. The growth rate of the capital stock and output is equal to the rate of population growth, or, if there is labor-saving (i.e. Harrod-neutral) technological progress, by the sum

---

[21] An illuminating discussion of the Harrod-Domar growth model can be found in Dixit (1976).

of the rate of technological change and the population growth rate. That is, the ratio of the capital stock to effective labor services is constant in steady state. What are the distributional implications of the Solow-Swan model? Since the interest rate is constant and wages grow at the exogenous rate of technological progress (both the interest rate and real wages depend on the ratio of the capital stock to effective labor services), the model predicts a stable long-run functional income distribution, i.e. capital and labor income grow at the same rate as the capital stock and output. Similar steady state dynamics hold in refined versions of the model of Ramsey (1928) [see Cass (1965) and Koopmans (1965)]. In these neoclassical growth models, savings and thus capital accumulation are endogenously derived by household optimization, whereas the Solow-Swan model treats the saving rate as exogenous.

In fact, in the mid of the 20th century, a stable functional income distribution has been viewed as a stylized fact. As Kaldor (1956, p. 84), who aimed to explain a stable functional income distribution as well, pointed out:

> In fact no hypothesis as regards the forces determining distributive shares could be intellectually satisfying unless in accounting for the relative stability of these shares in the advanced capitalist economies over the last 100 years or so, despite the phenomenal changes in the techniques of production, in the accumulation of capital relative to labour and the real income per head.

Whether or not this view is still supported by the empirical evidence will be discussed in section 1.5.

1.3 The Kuznets hypothesis and some cross-section data

Rather than with the *functional* income distribution, Kuznets (1955) was concerned with the evolution of the *personal* income distribution as the economy grows. According to his famous hypothesis, over time inequality increases at low development stages and decreases at high development stages. Kuznets suggested such an inverted U-shaped relationship between the development of an economy and income inequality, known as the *Kuznets curve*, for the following reasons. In earlier phases of development, inequality would rise due to industrialization and urbanization. Viewing the total income distribution as combination of the income distribution of the rural and the urban population, Kuznets has shown in a numerical example that if income of the rural population is both lower on average and more equally distributed than in urban areas, the income share of the poorest 20 percent of the population declines as the proportion of the urban population goes up. Moreover, Kuznets argued that the per capita income differential between rural and urban areas would persist or even widen as the economy grows, since productivity growth would not be slower in urban than in rural areas.[22]

> The relative difference in per capita income between the rural and urban populations does not necessarily drift downward in the process of economic growth: indeed, there is some evidence to suggest that it is stable at best, and tends to widen because per capita productivity in urban pursuits increases more rapidly than in agriculture. [Kuznets (1955, p. 8)].

According to his numerical example, the higher this income differential is, the more marked is the fall in the income share of the bottom quintile in the process of urbanization. However, Kuznets believed that inequality would eventually be driven down. First, he argued that if individuals with high abilities are also the ones with high income levels, there would be no reason that their descendants would be talented enough to have high earnings as well. (However, as will be seen below, models assuming imperfect capital markets and bequeathing imply that this view may be too optimistic). Second, Kuznets pointed out that

> the very fact that an increasing proportion of the urban population was [...] more able to take advantage of the possibilities of city life in preparation for the economic struggle meant a chance for organization and adoption, a better basis for securing greater income shares than was possible for the newly 'immigrant' population coming from the countryside or from abroad. [Kuznets (1955, p. 17)].

Third, Kuznets was aware of an increasing demand for redistribution as the economy grows, pointing out that

> in democratic societies the growing political power of the urban lower-income groups led to a variety of protective and supporting legislation, much of it aimed to counteract the worst effects of rapid industrialization and

[22] In a recent but already standard textbook about economic growth, Aghion and Howitt (1998) claim that Kuznets (1955) suggested an eventual reduction in income inequality in the growth process due to an "increasing scarcity of agricultural workers", implying that "wages in the rural sector are catching up with wages in the industrial sector" (p. 280). Such a mechanism has indeed been formalized by Bouguignon (1990). However, as the quotation above reveals, this is not what Kuznets suggested. If anything, Kuznets expected *widening* income differentials between the urban and the rural population over time.

> urbanization and to support the claims of the broad masses for more adequate shares of a growing income of the country. [Kuznets (1955, p. 17)].

In particular, he referred to the still valid fact that the marginal propensity to save is an increasing function in income which would tend to increase both wealth and income inequality as the economy grows.[23] Kuznets suggested that this development would lead to an increasing political pressure to redistribute wealth.

> One group of factors countering the cumulative effect of concentration of savings upon upper-income shares is legislative interference and 'political' decisions [...] aimed at limiting the accumulation of property directly through inheritance taxes and other explicit capital levies. [...] All these interventions [...] do reflect the view of society on the long-term utility of wide income inequalities. [...] This should be borne in mind in connection with changes in this view even in developed countries, which result from the process of growth and constitute a re-evaluation of the need for income inequalities as a source of savings for economic growth. [Kuznets (1955, p. 9)].

Whether or not low inequality and redistribution depress growth is a central question explored in the next section.

However, before going into these debates, one should briefly look at the evidence of an inverted U-shaped relationship between inequality and development of an economy (measured as per capita income). To date, economists have not reached a consensus whether or not such a relationship exists. However, it should be noted that tests of the original Kuznets hypothesis have been restricted by the insufficient availability of longitudinal data of inequality measures. Rather, the Kuznets hypothesis has been tested by using cross-section data. The results have been mixed.[24] Using new cross-section data by Deininger and Squire (1996) and the World Bank (1998), Fig. 1.2 shows that, if anything, the relationship between the Gini coefficient and real GDP per capita (in purchasing power parity) is negative rather than inverted U-shaped.[25]

---

[23] E.g. Browning and Lusardi (1996). According to Bosworth, Burtless and Sabelhaus (1991), average savings of the poorest two income quintiles in the U.S. are even negative.

[24] For supporting evidence, see e.g. Ahluwalia (1976) and Campanao and Salvatore (1988) who regress income shares of the bottom 20, 40 or 60 percent of the population on real GNP per capita and its square for a cross-section of countries.

[25] Income data stem from an updated version of the Penn World Tables (PWT) by Summers and Heston (1991). Moreover, only the high-quality subset of Deininger and Squire (1996) is included in the sample. If more than one Gini observation is available for a country, the latest one is taken. If the same latest observation for a

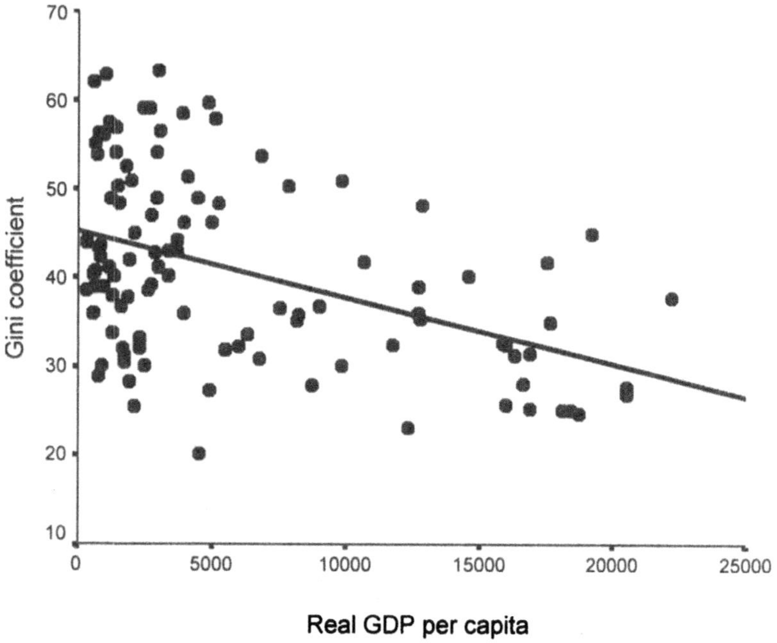

Figure 1.2: Income inequality and per capita income in 107 countries.
Data sources: GDP: PWT 5.6. Gini: Deininger-Squire (1996), World Bank (1998).

This is confirmed by estimating the equation

(1.1) $$GINI = \alpha + \beta_1 GDP + \beta_2 GDP^2 + \beta_3 GDP^3 + \gamma \, Dummy$$

with OLS. Including GDP only in the regression (i.e. imposing the restriction $\beta_2 = \beta_3 = \gamma = 0$) gives a negative coefficient β_1 which is significantly different from zero at the 1% level (adjusted $R^2 = 0.183$). However, all β-coefficients are insignificant when estimating a quadratic or cubic form or when dummies for Central/Latin American and African countries are included.[26] The same is true for estimating any equation for a

country is available in the two data sets [Deininger and Squire (1996), World Bank (1998)], the World Bank data is taken. Differences between the two data sources mainly arise for Scandinavian countries for which Deininger and Squire (1996) report unusually high Gini coefficients.

[26] Both continent dummies are negative and statistically different from zero at the 1% level.

subset of countries with income below U.S.$ 5000 (doing so, the number of observations drop from 107 to 67).

Another argument against the Kuznets hypothesis in cross-sectional data can be derived from Fig. 1.3. For the 19 OECD countries for which comparable Gini coefficients are reported by Gottschalk and Smeeding (1997, Fig. 2), the relationship between inequality and income is U-shaped rather than inverted U-shaped (with coefficients statistically significant from zero at the 5 % level; adjusted $R^2 = 0.201$).

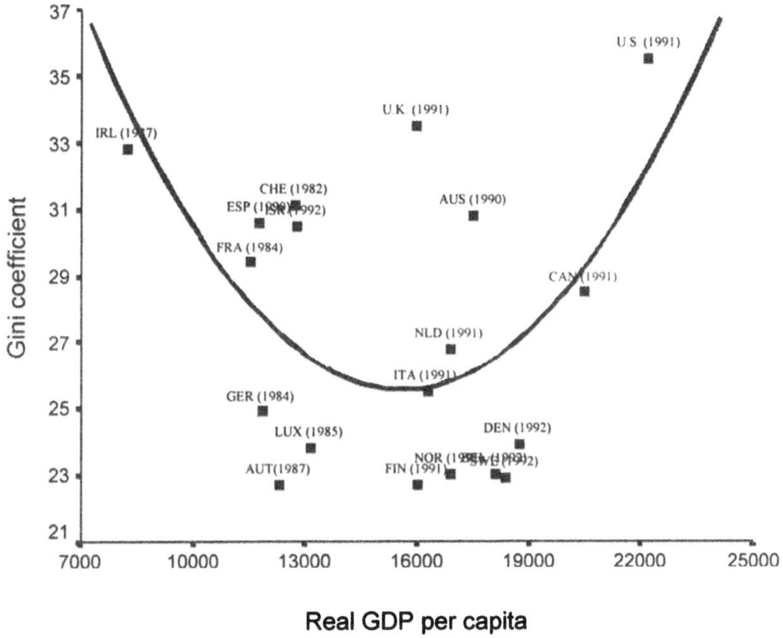

Figure 1.3: Income inequality and per capita income in 19 OECD countries. *Data sources:* GDP: PWT 5.6. Gini: Gottschalk-Smeeding (1997, Fig. 2).

However, the use of cross-section data may not be appropriate.[27] Taking into account that the original Kuznets hypothesis addresses an intertemporal relationship, Deininger and Squire (1998) exploit the panel data nature of the Gini observations which the authors describe in Deininger and Squire

---

[27] See e.g. Saith (1983, p. 382) who states: "As a paradigm, the cross-country U-hypothesis is arguably more of a hindrance than an aid to our comprehension of the relationship between economic growth and income distribution. It obfuscates more than it clarifies."

(1996). Doing so, they find an 'inverted U' which, however, is not robust. For instance, it vanishes when a dummy for Latin American countries is added. For Japan, Minami (1998) finds growing income inequality from the 1890s to the 1930s (estimating the pre-war income distribution at various points in time), sharply declining inequality in the years immediately after the Second World War, and a pretty stable income distribution thereafter. However, he does not view his findings as evidence for the Kuznets hypothesis since inequality did not continue to fall from the 1950s on and the decline in inequality has been due to ''post-war hyper-inflation and the new government policy of heavy taxes on assets and wealth'' (Minami (1998, p. 55). However, it is frequently neglected that one reason why Kuznets himself expected that income inequality would eventually decline as the economy grows was in fact that he expected an eventual increase in the level of redistribution.
Whereas Kuznets focused on the evolution of the personal income inequality and redistribution in a growing economy over time, the next section discusses if and how the level of both income/wealth inequality and redistribution matters for the growth rate of an economy.

1.4 How inequality and redistribution affects growth

Economists have long focused on the relationship between income distribution and efficiency, rather than on the relationship between inequality and growth.[28] For example, in public finance theory Musgrave and Musgrave (1976) emphasized both efficiency goals and the distribution function of the state. A tradeoff between efficiency and equality has been the center of the debate in the theory of optimal income taxation [see the pioneering work of Mirlees (1971)].[29] The view that countries face a trade-off between equality and efficiency has been firmly expressed by Okun (1975) who states:

> The contrasts among American families in living standards and in material wealth reflect a system of rewards and penalties that is intended to encourage

---

[28] Representative agent models in which a change in economic policy (e.g. lowering capital income taxation) promotes growth seem to suggest that also efficiency is raised [e.g. Lucas (1990), Rebelo (1991)]. For a critical discussion of this view, see e.g. Putterman, Roemer and Silvestre (1998).

[29] In this theory, agents differ in abilities and income taxation distorts their labor/effort supply decision. [See Atkinson and Stiglitz (1980, ch. 13) for an illuminating discussion].

> effort and channel it into socially productive activity. To the extent that the system succeeds, it generates an efficient economy. But that pursuit of efficiency necessarily creates inequalities. And hence society faces a tradeoff between equality and efficiency. [...] It is, in my view, our biggest socioeconomic tradeoff, and it plagues us in dozens of dimensions of social policy.

In contrast, macroeconomic theory did not address the personal income distribution in the 1970s and 1980s. In particular, the relationship between inequality and growth has not been examined before the early 1990s when a negative impact of income inequality on the rate of growth has been observed (using cross-country regression analysis).[30] Growth models with heterogeneous agents addressing this relationship (like the models presented in chapters 2-4) belong to the so-called 'new growth theory' in which growth is either driven by (positive external effects of) investment in physical and human capital or by endogenous technological change.[31]
Roughly spoken, the notion of an equality-growth tradeoff still dominates the recent economic policy debate in many developed countries.[32] For instance, policy makers in Germany frequently claim that lowering marginal tax rates for both high income earners and owners of capital would encourage investment and growth. In examining the impact of inequality and redistribution on growth theoretically as well as empirically, this section also deals with the question if such proposals are supported by recent research. Moreover, other links between inequality and growth which are not related to redistribution are briefly discussed. Last but not least, the review of this literature motivates the analysis in chapters 2 and 3.

---

[30] E.g. Persson and Tabellini (1994), Alesina and Perotti (1994), Perotti (1996). Deininger and Squire (1996) cast doubt on these findings, using high-quality observations of their new data set of income inequality measures. However, using the same data set but a different definition of high-quality observations, Figini (1999) confirms the relationship.

[31] The consideration of heterogeneous agents in endogenous growth models is a very recent development. For instance, in their well-known book on growth theory, Barro and Sala-i-Martin (1995) make no reference to other than representative agent models.

[32] It is important to note that an equality-growth trade-off is generally very different from an equality-efficiency trade-off. For example, excess burdens due to (redistributive) taxation, by definition, accrue at an individual (i.e. microeconomic) level (i.e. each individual would prefer to pay the same amount of taxes in a lump-sum fashion, and only in this sense, redistribution reduces efficiency). Excess burdens are associated to substitution effects, whereas growth effects are total (i.e. macroeconomic) effects.

1.4.1 The politico-economic approach

Without considering growth, Meltzer and Richard (1981) have shown that income inequality, measured by ratio of median to average income, leads to a higher linear income tax rate which is redistributed lump-sum among the individuals in a voting model. In this model, given the specific assumption of a linear income tax, the median income earner is also the median voter who is pivotal under the majority voting rule.[33] Thus, the (relatively) poorer the median voter is, the higher is his/her preferred level of redistribution.[34] Bertola (1993), Persson and Tabellini (1994) and Alesina and Rodrik (1994) take up this idea in models with investment-driven growth. In these models, endogenous growth arises because of a positive externality of capital accumulation, yielding socially non-decreasing returns to capital. Under the assumptions of perfect capital markets and infinitely living consumers, a capital-poor median voter demands a high level of redistributive capital income taxation which in turn depresses the rate of economic growth. The distribution of the (physical or human) capital endowment is exogenous and some inequality remains even after redistribution. This is because voters take the adverse growth effects of capital income taxation into account. In Bertola (1993), redistribution raises the level of wages and lowers the return to capital, and the growth rate of wages depends on the rate of capital accumulation. In Persson and Tabellini (1994) capital income is proportionally taxed and redistributed lump-sum to the individuals. Alesina and Rodrik (1994) consider a model in which capital income taxes are levied to finance productive public investment expenditures. Thus, in their model taxation of capital may even enhance growth, but only for sufficiently small capital income tax rates. Taxation has an indirect redistributive effect on the level of wages which depends on the quality of the publicly provided infrastructure.[35]

---

[33] For a discussion of majority voting models of income taxation, see Romer (1975) and Roberts (1977).

[34] For further discussion of this idea, see Mueller (1989).

[35] Also Saint-Paul and Verdier (1992) and Glomm and Ravikumar (1993) analyze endogenous growth models with heterogeneous agents and majority voting. In Glomm and Ravikumar (1993), agents vote over the level of private versus public education, where schooling is the engine of growth. In contrast to private schooling, public schooling unambiguously leads to declining income inequality over time, whereas per capita income is generally higher under private education. If the majority of individuals has less than per capita income, schooling is shown to be publicly financed in voting equilibrium. Saint-Paul and Verdier (1992) show that the (tax-financed) level of public schooling (which has a redistributive effect)

Politico-economic models linking inequality to redistribution can only be analyzed under given specifications of the tax scheme. But the choice of the tax scheme determines whether or not redistribution, i.e. a reduction in inequality, is harmful for growth. Thus, by assuming growth-reducing taxation to be voted on, the result of the politico-economic growth models outlined above is not at all surprising in light of the result of Meltzer and Richard (1981).[36] Moreover, it has been criticized that the result critically depends on the assumptions of perfect capital markets and infinitely living agents. Relaxing the infinite horizon assumption, Bertola (1996) and Uhlig and Yanagawa (1996) show that shifting taxes from labor income to (human or physical) capital income (holding the tax revenue share in aggregate income constant) may even increase growth. [They consider a tax scheme similar to the one in Bertola (1993)]. The reason is that, in the absence of bequests, newly born individuals have to rely on labor income in order to build a capital stock, i.e. capital income solely accrues to the old. Thus, taxing capital income more heavily and cutting the tax rate on labor income leaves young agents with more income out of which to save, such that the overall impact on capital accumulation may indeed be positive. In contrast, the politico-economic models reviewed above imply that redistributing income from owners of capital to workers unambiguously slows down growth. The other critical assumption in investment-driven growth models concerns the role of borrowing constraints to finance certain projects. For instance, due to a moral hazard problem, individuals do not have access to credit markets, and thus are not able to finance higher education.[37] As will be seen in the following, by introducing imperfect capital markets in investment-driven growth models, the trade-off between redistribution and growth may vanish.

1.4.2 Capital market imperfections and non-convex technologies

In the last decade there has been a series of prominent papers, indicating that redistribution of wealth towards the poor (i.e. a reduction in both wealth and income inequality) may have a growth-promoting effect if capital markets are imperfect. Moreover, these models address the question whether or not inequality declines in the process of development, i.e. in the process of

---

rises with income inequality in voting equilibrium, implying that the relationship between inequality and growth in democracies does not have to be negative.

[36] See e.g. Roemer (1998) for a critique of this approach.

[37] Primary education is compulsory and free in many countries. Even though higher education may also be free in some countries, it is never compulsory (i.e. a decision variable of households) and comes at the cost of foregone consumption.

endogenous wealth accumulation, as suggested by Kuznets (1955). (Models in which inequality eventually declines as the economy grows are called 'trickle down growth models'). Most models assume that a fixed capital outlay is necessary for an individual to start a production activity [Banerjee and Newman (1993), Galor and Zeira (1993), Perotti (1993), Aghion and Bolton (1997), Matsuyama (1998)], e.g. to open a firm or to finance secondary schooling.[38] However, Bénabou (1996) and Piketty (1997) have shown that this assumption is not necessary to obtain the result that inequality may be harmful for growth if capital markets are imperfect. The latter result also turns out not to be sensible to the source of capital market imperfection. Aghion and Bolton (1997) and Piketty (1997) assume that effort (which is individually costly, but positively affects the probability that the investment will yield a positive return) is not observable for lenders and a borrower's repayment to his/her lender cannot exceed his/her future income. This gives rise to a moral hazard problem, implying that poor individuals are credit-constrained.[39] Thus, redistribution may stimulate total investment since more individuals become able to borrow.[40] In both models inequality may eventually decline as the economy grows. Whereas the economy of Aghion and Bolton (1997) converges to a unique steady state wealth distribution (if it converges at all), there may be multiple steady states in Piketty (1997). In the latter case a one-shot redistribution may help to bring the economy from a steady state with slow accumulation and high wealth inequality to a one with fast accumulation and low wealth inequality. A similar result is obtained in both Banerjee and Newman (1993) and Matsuyama (1998) who assume that there is a repayment enforcement problem which leads to an upper limit on the amount of individual borrowing. Thus, redistribution may help individuals to finance the indivisible investment requirement. In Galor and Zeira (1993) capital markets are imperfect in the sense that the rate at which individuals can

---

[38] In an interesting model, Jovanovic (1998) shows that the assumption of non-convex technologies necessarily implies that the initial income (i.e. wage) distribution is unequal. In contrast, in the models reviewed in this section, the initial (unequal) income distribution is exogenous.

[39] Sappington (1983). See also the seminal paper on credit-rationing by Stiglitz and Weiss (1981).

[40] In Aghion and Bolton (1997), also effort provision may increase after redistribution. This is because wealthier net borrowers (who are not credit-constrained) have to borrow less in order to finance the set-up costs, and thus have to repay less if the project succeeds. (If the project fails, they cannot pay back anything). In other words, the expected marginal returns (and thus effort levels) are higher for richer borrowers who have to share a lower fraction of their marginal returns with lenders.

borrow exceeds the lending interest rate. Due to the high costs of borrowing, an investment in human capital is only optimal for individuals with sufficiently high amounts of inherited wealth. In the long-run, if the wedge between the borrowing and the lending rate is sufficiently large, some dynasties will always invest whereas others will never invest to become skilled workers. The authors also show that some descendants of skilled workers will eventually become unable to invest in human capital. In this sense, inequality increases as the economy grows.[41] Perotti (1993) and Bénabou (1996) assume that a loan market simply does not exist, again implying that redistribution may foster growth. Moreover, both authors derive a voting equilibrium with respect to a given tax-transfer scheme. That is, they assume in a politico-economic growth model that capital markets are imperfect.[42] The crucial assumption in Bénabou (1996) is that individual marginal returns to education are decreasing. Since the additional output of a marginal increase in human capital for a poor individual exceeds the output loss of the reduced after-tax income of a rich individual, a tax increase may be growth-enhancing even though the technology is convex (i.e. there is no fixed investment necessary to work as skilled). Moreover, unlike the models discussed in the preceding subsection, the models of both Perotti (1993) and

---

[41] See Atkinson (1997) for a further discussion and suggestions of similar mechanisms. Galor and Moav (1999) suggest a unifying approach to the following two conflicting viewpoints. On the one hand, given the stylized fact that the individual marginal propensity to save is rising in income, inequality fosters growth due to a higher accumulation rate of physical capital. But on the other hand, as discussed in this section, inequality slows down human capital accumulation rates due to borrowing constraints. In their model, in early stages of development physical capital accumulation is the prime engine of growth, thus supporting the first view, but at later stages of development human capital accumulation becomes the prime engine of growth, supporting the second view. In their model, human capital becomes increasingly important for growth due to a capital-skill complementarity [see Goldin and Katz (1998) and section 1.5 for further discussion]. Moreover, since human capital has a very different nature than physical capital as the former is embodied in humans, widely-spread human capital (i.e. more equality) in a society may eventually foster growth when the physical capital stock is sufficiently large.

[42] The reason why these contributions are quoted here rather than in the preceding subsection (they fit into both categories) is that the results are more consistent with the other models of imperfect capital markets than the results of the politico-economic models discussed in subsection 1.4.1. See subsection 1.4.5 for a discussion of the empirical discrimination between the different results and approaches.

Bénabou (1996) imply that inequality may be negatively related to redistribution in voting equilibrium.[43]

1.4.3 Sociopolitical instability

According to another widely recognized link between inequality and growth, an unequal income/ wealth distribution gives rise to sociopolitical conflicts which, in turn, reduce the rate of accumulation and thus growth.[44]
Benhabib and Rustichini (1996) consider a dynamic game in which organized social groups (i.e. the players) can capture a larger share of output by expropriation of other groups (i.e. through direct appropriation or manipulation of the political system). Rent-seeking reduces the resources which are left for capital accumulation, in turn depressing growth. In the model, social instability (i.e. defection of all groups in a prisoners' dilemma situation) hurts poor groups less than rich ones relative to the initial gain of

[43] In the non-overlapping generations model of Perotti (1993), each individual lives two periods and a linear endowment tax is used to finance lump-sum transfers. The share of individuals who are able to finance the fixed human capital investment depends on the income distribution in the first period after taxes and transfers have been made. There are three income groups, and the decisive voter is in the middle class. Total investment affects the second-period income of all individuals due to a spill-over effect. If per capita income is high (i.e., all other things equal, total investment is high), and inequality is low in the sense that the low and middle endowment are not too different, this may give the median voter a motive to vote for a tax rate which is sufficiently high to enable also the poor to invest. As this motive vanishes when inequality is high, there is a negative relationship between inequality and growth in rich economies. However, in poor economies, the model predicts a positive relationship between inequality and growth, where the endowment distribution is said to be more equal the less the middle and high endowment differs. With a fairly equal endowment distribution, middle class individuals have a motive to vote for high taxes in order to increase income in the first period, since the high income group would only be able to invest with very low taxes. In contrast, if inequality is high, taxes are low in voting equilibrium, i.e. the high income group is kept able to invest. Thus, the relationship between initial inequality and investment is positive rather than negative. Also Bénabou (1996) shows that the tax rate (and thus redistribution towards the poor) is not necessarily increasing with inequality. This holds if the pivotal voter is wealthier than the median, i.e. if there is a 'wealth bias' in the political system.

[44] The relationship between inequality and social instability has long been discussed among political scientists as well as economists [e.g. Hibbs (1973), Gupta (1977); see also Falkinger (1999) for a recent study and further references], although theoretical studies concerning the relationship between inequality and growth through this channel are rare.

expropriation. Thus, the poorer a group is, the more likely it is engaged in rent-seeking. In this sense, income inequality is positively related to expropriation, and thus negatively to growth.[45] Grossman and Kim (1996) model the behavior of two dynasties along a growth path: a potential predator and its prey. Besides allocating inherited wealth to productive capital, agents optimally choose the amount of resources devoted to predation and to the defense of predation, respectively. As long as current wealth (or initial wealth, respectively) of the potential predator is small relative to the prey dynasty, only some resources are devoted to defensive measures. However, as relative wealth of the predator dynasty increases due to predation, deterrence of predation becomes more costly so that the rate of accumulation slows down. In this sense, high inequality is associated with slow growth.[46] Palda (1999) points to the relationship between redistribution and property rights with respect to their impact on growth. According to his model, redistribution towards the poor may induce some predators (''wealth takers'', as he calls it) to become ''wealth makers'' which may lead to an increase in the rate of growth. It is shown that redistribution spurs growth only if property rights are sufficiently secure.

1.4.4 Other approaches

Finally, four less frequently explored channels are briefly discussed.
The first one concerns a link between income inequality and innovations of new consumer goods, which in turn affect growth [Zweimüller (1998)]. The basic idea is that under the assumption of hierarchical preferences, basic needs are satisfied first such that not all goods are demanded by all consumers. Since richer individuals demand a larger number of goods, income inequality affects the demand for new goods. Since research activities depend on the 'thickness' of a market, firms have less incentives to innovate if income inequality is high.
Second, Dasgupta (1993) suggests a channel between poverty and growth. If poverty is high, average food intake is low and nutrition is bad, which in turn affects both the level and the growth rate of labor productivity.

---

[45] Moreover, for a given income distribution, both a lower capital stock and a lower consumption level makes defection of an interest group more likely since the marginal utility of consumption is high and the deterrence of a depreciation of the capital stock due to social instability is low. Thus, the economy may get stuck in a 'poverty trap'.

[46] Whereas the identity of the exploiting and exploited group is exogenous in Grossman and Kim (1996), the identity of the exploiter is endogenously determined in the model of McDermott (1997).

Third, Corneo and Jeanne (1999a) and Knell (1999) examine the relationship between inequality and growth when individuals make social comparisons. In Corneo and Jeanne (1999a), *social status* of individuals depends on relative wealth levels. (See chapter 2 for a discussion of other status-seeking motives). The authors show that under incomplete information among individuals about wealth levels, there is an ambiguous relationship between inequality and growth. In contrast, this relationship is negative in Knell (1999). In his model, if young individuals have a higher concern for their relative standing than old individuals and if the reference group to which individuals socially compare themselves do not contain poorer individuals (i.e. social comparisons are not downward-oriented), then higher inequality depresses growth.[47]

Fourth, Perotti (1996) suggests a channel in which high inequality induces high fertility rates and high fertility depresses per capita growth through low aggregate investment in human capital.[48] Figini (1999) rather suggests that one should introduce borrowing constraints into a model with endogenous fertility and heterogeneous agents. According to his notion, poor individuals

---

[47] Again, see chapter 2 for a further discussion of these results.

[48] This notion is inspired by representative agent models of endogenous fertility [e.g. Becker and Barro (1988), Becker, Murphy and Tamura (1990)]. In these models, it is assumed that parents decide how much to consume, how many children to have and how much to invest in the human capital of their children. In other words, parents face a trade-off between the 'quantity' and 'quality' of children. All other things equal, the 'demand' for children rises with individual income. Moreover, income of an individual positively depends on the human capital investment the parents made in his/her favor. This may be called an income effect of higher human capital with respect to the demand for children. However, raising children has opportunity costs in form of foregone consumption while rearing a child. Thus, opportunity costs of child-rearing rise with the amount of human capital an individual has embodied. This is a substitution effect of higher human capital. Now suppose, the income effect of an increase in human capital dominates the substitution effect for *low* levels of human capital, and vice versa for *high* levels of human capital. In this case, poor households would have more children than rich households, whereas rich households would invest more in the human capital of their children. According to the notion of Perotti (1996), in this set-up, more equality (in the distribution of human capital) implies a decrease in aggregate fertility rates and an increase in enrollment ratios to schooling, thus fostering growth. Recently, Morand (1999) proposed a similar mechanism, however focusing on the role of children to provide old-age support to their parents [see e.g. Dasgupta (1993) for a discussion]. It is suggested in this paper, that there exists a threshold level of income above which parents increasingly substitute away from quantity of children towards human capital investments in children.

invest in the number rather than the quality of children since they can neither afford schooling for their children nor borrow for it.
Besides the mechanisms outlined in 1.4.1-1.4.3, we come back to both the fertility issue and the role of social comparisons in the next subsection, where the empirical results about the impact of inequality and redistribution on growth are discussed.

1.4.5 Empirical evaluation of the various approaches

Most empirical cross-country studies find that inequality is negatively related to growth, fitting a reduced-form equation like

$$(1.2) \qquad GROWTH = \beta_0 + \beta_1 INEQ + \gamma \boldsymbol{X},$$

where *GROWTH* is the average growth rate of GDP per capita over a certain time period (e.g. 20 years), *INEQ* is an (in)equality measure like the Gini coefficient or the percentage of income accruing from certain quintiles (e.g. the sum of the income share of the third and fourth quintile of the income distribution) and $\boldsymbol{X}$ is a vector of other control variables, known to potentially affect the growth rate, e.g. initial per capita income (in order to control for convergence in the growth rates as predicted by neoclassical growth theory) and the level of investment in human or physical capital.[49]
However, although most of the theories reviewed in this section indeed predict a negative inequality-growth relationship, the specific *channel* through which the growth rate is affected by inequality cannot be identified with reduced-form estimates. As we have seen above, inequality may negatively affect the rate of growth due to the following reasons. First, according to a popular hypothesis, higher inequality leads to more redistribution, which in turn reduces growth (*fiscal policy approach*). Second, higher inequality should reduce human capital investments if capital markets are imperfect, thus depressing growth (*imperfect capital markets approach*). Third, higher inequality should induce political or social instability, harming growth (*sociopolitical instability approach*). Fourth, higher inequality should increase fertility rates (and decrease human capital investment), negatively affecting growth (*fertility approach*). Testing either channel can be achieved by simultaneously estimating the following two equations:

---

[49] For a tabular overview of the results of these kinds of empirical studies, see Bénabou (1996, Tab. 2), Aghion and Howitt (1998, Tab. 9.1), Figini (1999, Tab. 1).

$$(1.3) \qquad GROWTH = \beta_0 + \beta_1 CHANNEL + \gamma \boldsymbol{X},$$

$$(1.4) \qquad CHANNEL = \delta_0 + \delta_1 INEQ + \boldsymbol{\theta W},$$

where ***X***, ***W*** are other controls. Depending on the theoretical approach to be tested, *CHANNEL* denotes a measure of *redistribution policy* (e.g. average or marginal income tax rates, the ratio of tax revenue to GDP, the ratio of total government expenditure to GDP etc.), a measure of *investment in human capital* (e.g. the enrollment ratio in secondary schooling), a measure of *sociopolitical instability* (e.g. a weighted sum of political assassinations, violent deaths relative to population size and the number of successful and unsuccessful coups) or the (net) *fertility rate*. Note that, according to the theories reviewed above, these variables cannot just be included as additional controls in (1.2) since they are expected to be correlated with inequality, as recognized in (1.4). According to the fiscal policy approach, inequality should positively affect redistribution (i.e. $\delta_1 > 0$) and redistribution should decrease growth (i.e. $\beta_1 < 0$). Note that if the variable *CHANNEL* in (1.3) and (1.4) is specified as a redistribution measure, $\beta_1 > 0$ would be indirect evidence for the capital market imperfections approach, since the existence of borrowing constraints have been shown to give rise to growth-promoting effects of redistribution policies; $\delta_1 \leq 0$ would be consistent with the models of Perotti (1993) and Bénabou (1996) who both combine the fiscal policy approach and the capital market imperfections approach.

Joint estimation of (1.3) and (1.4) has been conducted by Perotti (1994, 1996), Alesina and Perotti (1996) and Figini (1999).[50] According to these studies, the fiscal policy approach is not at all confirmed. Not a single redistribution measure is found to be positively affected by inequality.[51] Moreover, policy variables do not negatively enter the growth equation (1.3) at a significant level. Using the marginal tax rate, social security/welfare spending and spending for health/housing as redistribution measures, Perotti (1996) even finds a positive and significant relationship between redistribution and growth. However, significance vanishes for some other redistribution measures. Estimating (1.3) only, rather than estimating (1.3)

[50] Perotti (1994) and Alesina and Perotti (1996) replace *GROWTH* in (1.3) with the investment to GDP ratio.

[51] Rather, Figini (1999) finds evidence for a reversed causality, i.e. inequality would be high because redistribution is low. Note that measured gross income inequality does not reflect inequality of market income since inequality data sets also contain pension and transfer payments.

and (1.4) jointly, Persson and Tabellini (1994) find a negative but not significant impact of the ratio of government transfers to GDP on the growth rate, whereas Sala-i-Martin (1996) finds a positive and significant impact of government transfers on growth. According to Easterly and Rebelo (1993), various average and marginal tax rates have negative, but generally not significant impacts on growth.

However, one should note that redistribution measures can take other forms than transfer payments, e.g. through minimum wage laws or collective bargaining. For example, according to a frequently made claim, unions try to reduce wage inequality in wage bargaining. But this is likely to be relevant in a few developed countries only. A more serious objection against the finding that fiscal policy is irrelevant for growth is made in a recent article by Kneller, Bleaney and Gemmell (1999). The authors show that inconclusive results concerning the relationship between fiscal policy variables and growth may be due to the negligence of the *fiscal budget constraint* in the estimated equations. That is, by including only *single* policy variables into a growth equation like (1.3), one does not account for the fact that tax rates and public expenditure levels are related. But this relationship potentially biases the estimated coefficients on taxation if some kind of public spending has positive growth effects.[52] Controlling for these effects separates out the growth-reducing effects of taxation. Observing the fiscal budget constraint, the authors find a *negative* and significant impact of taxation (i.e. the ratio of the tax revenue from income, profit, property, social security and payroll taxes to GDP) on growth in a sample of OECD countries.

In sum, whether or not taxation affects growth, inequality is not positively related to taxation or redistribution, respectively, unlike suggested by the fiscal policy approach. In contrast, strong evidence is found for the imperfect capital markets approach. Perotti (1994, 1996) and Figini (1999) find that inequality negatively affects school enrollment ratios of both men and women. Moreover, and frequently confirmed before, the school enrollment ratio positively affects growth.[53] Concerning the sociopolitical instability approach, results are somewhat more mixed.[54] Perotti (1994, 1996) and Alesina and Perotti (1996) find strong evidence for the notion that inequality leads to sociopolitical instability and thus to slow growth. In

[52] Obvious examples are public education [e.g. Saint-Paul and Verdier (1992), Glomm and Ravikumar (1993)] and infrastructure expenditures [e.g. Barro (1990)].

[53] E.g. Barro and Sala-i-Martin (1995).

[54] Palda (1999) criticizes that empirical studies addressing this channel neglect the role of property rights.

contrast, Figini (1999) does neither find a significant impact of inequality on his measure of political instability, nor that political instability is a strong impediment to growth (the coefficient β_1 is negative but not statistically significant). Since political instability significantly enters in the reduced-form equation (1.2), if used as additional control variable, Figini (1999) concludes that "social and political instability is an important variable in explaining growth but it seems to work independently from inequality rather than being a transmission to growth" (p. 35). Thus, Bénabou (1996) may be right to assert that "what really matters is not income inequality per se, but inequality in the relative distribution of earning and political power" (p. 18), although this notion is difficult to test empirically. Concerning the fertility approach, Perotti (1996) finds strong evidence for its validity. Figini (1999) also finds evidence for a positive impact of inequality on the fertility rate. Moreover, he finds that the ratio of M2 to GDP, introduced as (rather rudimentary) proxy for the development of financial markets, negatively enters the fertility equation, while leaving the inequality coefficient significant. This is consistent with his view that borrowing constraints affect fertility rates.[55]

Finally, the theory of Knell (1999) is concerned with a growth reducing effect of inequality due to social comparisons (*social comparisons approach*). As a direct test of this theory seems out of reach, Knell (1999) suggests that the negative relationship between inequality and growth in a reduced-form estimation of equation (1.2) should be strongest in rich countries. This is because status concerns would play a greater role in rich countries. Whereas he finds evidence for this hypothesis, Figini (1999) does not find a significant difference between samples of high-income and low-income countries (even when using the same cut-off income level to separate rich from poor countries as Knell).

1.4.6 Conclusion and outlook to chapters 2 and 3

So what have we learned from this section and what needs to be explored further? It has been shown that the growth rate of per capita income is negatively affected by inequality, and this relationship cannot be explained by the fiscal policy approach. This raises strong doubts on the widely accepted notion that high inequality and little redistribution through the tax system is needed to secure high growth. Empirical evidence rather supports theoretical results which link high inequality to low investment rates if

[55] However, it should be noted that R^2 does not improve by the conclusion of this variable as additional control. Moreover, fertility negatively affects growth, but only at the 10% level.

individuals face borrowing constraints. Moreover, but to a lesser degree, there is evidence for both the sociopolitical instability and the fertility approach. It has been argued that in view of imperfect capital markets, redistribution may help individuals to overcome borrowing constraints, thus fostering investment (in human and physical capital) and growth. In contrast, the theoretical result that inequality adversely affects growth due to higher demand for government spending in voting equilibrium is rejected by the data. Finally, one can conclude that the role of social comparisons for the relationship between inequality and growth certainly needs further exploration (although it will be hard to come up with a convincing empirical test).

This leads us to a brief outlook to chapters 2 and 3. If one accepts that the outlined models of imperfect capital markets and sociopolitical instability as well as the fertility approach are quite consistent with the empirical evidence, one may be satisfied with these lines of research. However, the questions how the inequality-growth relationship is affected by individual demands for public policy and by social comparisons are certainly still open to debate.

Chapter 2 looks at the relationship between individual concerns for relative consumption levels, inequality and growth. On the one hand, like in Knell (1999), the role of social comparisons (i.e. status concerns) for the relationship between inequality and growth is explored. On the other hand, the impact of status concerns on growth for a given income distribution is analyzed; first, in examining the impact of status concerns on saving behavior, and second, through the politico-economic channel. Regarding the second link, concerns for relative standing affect the voting outcome with respect to income redistribution and thus indirectly the incentives to save.[56]

Chapter 3 addresses the empirical falsification of the fiscal policy approach, i.e. the fact that inequality does not affect growth through the demand for redistributive taxation. As outlined above, both Perotti (1993) and Bénabou (1996) can explain this result in their voting models by departing from the fiscal policy approach in the following way: capital markets are assumed not to be perfect and the median voter not necessarily to be pivotal.[57] In contrast, the more differentiated view of the inequality-growth relationship with

---

[56] Note that, in contrast to the fiscal policy approach, not a change in inequality affects the voting equilibrium, but status concerns for a given distribution of income.

[57] Both Perotti (1993) and Bénabou (1996) relax the assumption of perfect capital markets and, in addition, Bénabou (1996) departs from the hypothesis of a decisive median voter in deriving a non-linear relationship between inequality and redistribution.

respect to the politico-economic channel obtained in chapter 3 is due to the following modification of standard politico-economic growth models. Rather than considering financial transfers towards the median income earner, which are barely observed in most economies, redistribution takes place through the public provision of consumption goods and services. (The assumptions of proportional income taxes, infinitely living agents and perfect capital markets are maintained). In chapter 3 also some cross-section evidence for the hypothesis that income inequality does not positively affect the level of public consumption is provided.

1.5 How technological change affects the income distribution

As we have seen in the preceding section, inequality has mainly been suggested to affect growth through its impact on investment in physical or human capital rather than through technological change.[58] However, technological changes have long been suggested to affect the income distribution, especially through their impact on earning opportunities in the labor market (see section 1.1). This channel has recently received tremendous attention due to the steadily increasing wage inequality and/or unemployment of low-skilled labor in developed countries in the last decades. This section reviews the theoretical as well as the empirical literature on the impact of technological change on the demand for high-skilled relative to low-skilled labor. Especially, the popular hypothesis of *skill-biased technological change* (SBTC) is discussed. According to this hypothesis, new technologies cause a shift in the production function which contains high-skilled and low-skilled labor as arguments, such that the relative marginal productivity rises. As a consequence, relative labor demand shifts in favor of high-skilled labor, and, if wages are sufficiently flexible, relative wages of high-skilled workers increase for a given relative labor supply.[59]

Fig. 1.4 gives a simple representation of this view. Suppose there is a perfectly competitive and segmented labor market for skilled and unskilled

---

[58] An exception is the approach of Zweimüller (1998) who suggests an effect of the income distribution on the incentives of firms to innovate (see section 1.4.4).

[59] There is an ongoing debate whether changing earnings inequality is caused by the liberalization of trade between developed and developing countries or by SBTC. Although these issues are far from being settled, trade-related channels are not considered in the following. Rather, the focus is on technological change. For a discussion of the recent debate about technology bias versus trade, see e.g. Aghion, Caroli and García-Peñalosa (1999), Beißinger and Möller (1999), Franz (1999).

labor. Relative labor demand is falling in the relative wage of skilled labor and, for simplicity, relative supply of skilled labor is completely inelastic. With a relative labor demand shift only, the economy moves from an initial equilibrium given by the intersection point I to an equilibrium given by the intersection point II. (We come back to Fig. 1.4 later).

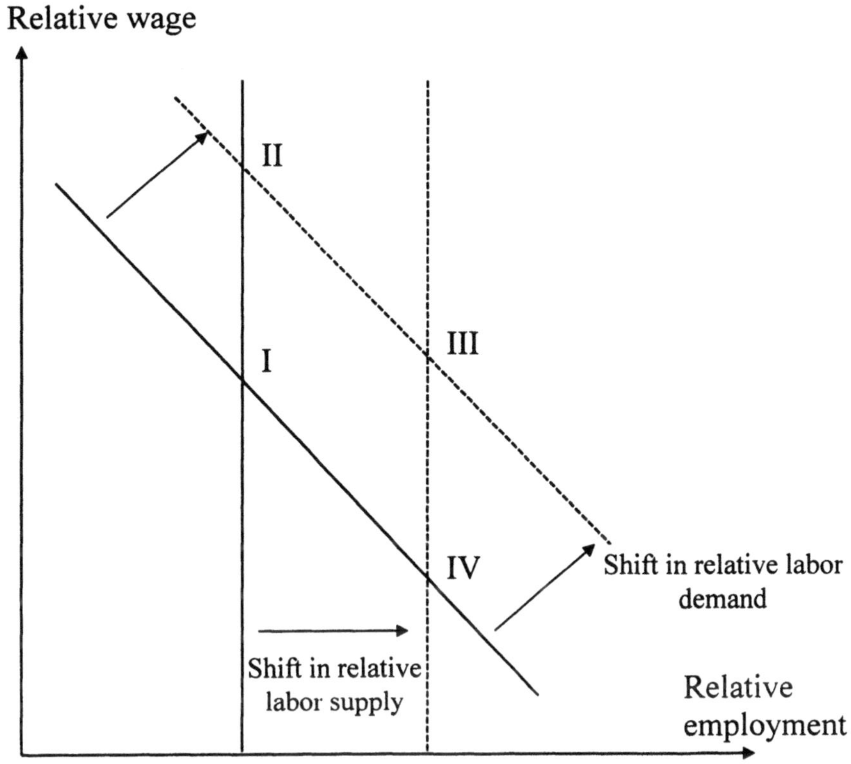

Figure 1.4: High-skilled relative to low-skilled employment and relative wages with changing relative labor demand and supply.

According to another suggestion, also capital accumulation may change the relative marginal productivity in favor of qualified workers, if physical and human capital are technological complements.[60] Moreover, as seen in sections 1.1 and 1.2, there has been a long debate about the effects of capital

[60] For theoretical models on the capital-skill complementarity, see e.g. Galor and Tsiddon (1997) and Galor and Moav (1999); for empirical evidence, see e.g. Bergström and Panas (1992) and Goldin and Katz (1998).

accumulation on the functional income distribution (although this issue has been neglected by economists in the last decades). However, in line with the recent discussion, this section focuses on technological change rather than capital accumulation.

Section 1.5 is organized as follows. In 1.5.1, recent trends of both earnings inequality and income inequality in the OECD are presented. Although the recent literature primarily focuses on the personal earnings distribution, we also look at the functional income distribution, i.e. the labor and non-labor share in national income, respectively. These trends are supposed to motivate the recent debate about the relationship between the dramatic changes in labor market outcomes which have occurred in some countries and changes in production technologies. This debate is reflected in 1.5.2. On the one hand, the literature on endogenous SBTC change which has emerged to explain recent trends in earnings inequality is reviewed. On the other hand, the empirical evidence supporting the view that technological change has played an important role for the observed changes in labor market outcomes is discussed. Finally, 1.5.3 summarizes the discussion and outlines some open questions which are addressed in chapter 4.

1.5.1 Earnings and income inequality in the OECD: empirical evidence

In the last two decades, overall wage inequality in the OECD has increased sharply in the U.S. and Great Britain, but, if at all, to a lesser degree in other OECD countries.

According to Tab. 1.1, between 1980 and 1993 the ninetieth to fiftieth percentile ratio of the *earnings distribution*, i.e. the wage level exceeded by exactly 10 percent of employees relative to the median wage, increased sharply in the U.S., Great Britain and in Italy. The fifties to tenth percentile ratio, i.e. median earnings relative to the wage level not exceeded by 10 percent of employees, also increased sharply in the U.S. and Great Britain, but remained largely unchanged in other countries (even somewhat falling in Germany).[61] For most countries, these numbers are consistent with the changes in overall *income dispersion*, measured by Gini coefficients, as reported in Tab. 1.2.

Whereas Gini coefficients increased in Australia, the U.S. and Great Britain, it showed no clear trend in most other countries. By decomposing measures of the income distribution into different sources of income for both Great

---

[61] Juhn, Murphy and Pierce (1993) find that real average weekly wages of the least skilled workers, measured by the bottom percentile of the wage distribution, even fell by about 5 percent between 1963 and 1989, whereas they have risen by about 40 percent for the top percentile.

Britain and Germany, Giles, Gosling, Laisney and Geib (1998) find that both levels and changes (between 1984 and 1992) of overall *income* inequality in both countries can almost solely be attributed to levels and trends of *earnings* inequality, respectively.

		1980	**1983**	**1988**	**1993**
Australia	D9/D5	1.70	1.68	1.72	1.76
	D5/D1	1.60	1.64	1.68	1.64
Canada	D9/D5	1.67 (1981)	n.a.	1,71	1.73
	D5/D1	2.07 (1981)	n.a.	2.23	2.19
France	D9/D5	2.03	2.06	2.12	2.13
	D5/D1	1.66	1.62	1.62	1.61
Germany	D9/D5	n.a.	1.63	1.65	1.64
	D5/D1	n.a.	1.46	1.42	1.37
Great Britain	D9/D5	1.62	1.70	1.78	1.84
	D5/D1	1.55	1.60	1.70	1.74
Italy	D9/D5	1.43	1.46	1.56 (1989)	1.65 (1994)
	D5/D1	1.63	1.53	1.39 (1989)	1.60 (1994)
Japan	D9/D5	1.63	1.65	1.70	1.71
	D5/D1	1.60	1.64	1.64	1.61
Sweden	D9/D5	1.61	1.55	1.57	1.62
	D5/D1	1.31	1.30	1.34	1.36
U.S.	D9/D5	1.76	1.86	1.99	2.00
	D5/D1	1.85	1.98	2.05	2.06

Table 1.1: Trends in OECD wage dispersion (males).
Source: OECD (1996, Tab. 3.1).
Notes: D9, D5, D1 refer to the ninetieth, fiftieth and tenth percentile level of earnings; e.g. 10 percent of employees earn less than D1 earnings. Levels of the inequality indicators, i.e. D9/D5 and D5/D1, are not comparable across countries due to different methods of measurement. n.a. = not available.

However, given the changes in the Canadian wage dispersion reported in Tab. 1.1, it is somewhat puzzling that, according to Tab. 1.2, income inequality decreased in Canada between 1985 and 1991.[62]

---

[62] By making inequality measures comparable across countries, Gottschalk and Smeeding (1997) show that not only *changes* in overall wage and income dispersion but also its *level* substantially differs among countries. For example, they find that the ninetieth to tenth *earnings* percentile ratio for full-year, full-time workers in the U.S. in 1991 has been more than twice as high than that of Germany

	Mid 1970s	Mid 1980s	Early 1990s
Australia	34.33 (1976)	37.58 (1985)	41.72 (1990)
Canada	31.62 (1975)	32.81 (1985)	27.65 (1991)
France	34.85 (1979)	34.91 (1984)	n.a.
Germany	30.62 (1973)	32.20 (1984)	n.a.
Great Britain	23.30 (1975)	27.10 (1985)	32.40 (1991)
Italy	35.98 (1978)	33.58 (1986)	32.19 (1991)
Japan	n.a.	35.90 (1985)	35.00 (1990)
Sweden	33.12 (1976)	31.24 (1985)	32.44 (1992)
U.S.	34.42 (1975)	37.26 (1985)	37.94 (1991)

Table 1.2: Overall income dispersion in OECD countries measured by Gini coefficients.
Source: High quality data set by Deininger and Squire (1996).
Notes: Within a country, all observations are from the same source (all sources are reported in the used data set). If possible, data are from 1975, 1985 and the latest observation in 1990s. If observation from 1975 (1985) is not available, the closest observation to 1975 (1985) is reported (year of observation in parenthesis). n.a. = not available.

Changes in overall wage dispersion can be decomposed in changes in wage dispersion *across* and *within* groups. For instance, employees are grouped according to their education (e.g. years of schooling, 'university degree', 'high-school degree', 'less than high-school'), occupation (e.g. 'white-collar', 'blue-collar'), experience, gender, race, age, and so on. Regressing (the log of hourly) wages on these and other personal characteristics, a famous procedure suggested by Mincer (1974) for identifying the average returns to schooling (as common, such an estimated equation is referred to as 'Mincer-equation' in what follows) reveals how much of the wage

---

in 1984 and that of the Netherlands in 1987. Moreover, whereas the tenth percentile in the U.S. has earned a wage about merely one third than the national median in 1991, and only slightly more in Canada 1987, in Germany 1984 the tenth percentile has earned almost two thirds of the median wage. (See their Tab. 1). Similarly, in 1991, the tenth percentile of the *income* distribution in the U.S. and Great Britain had a disposable income of just 36 percent and 44 percent of the median disposable income, respectively. (See their Fig. 2). In contrast, the same figure is in the range of 55-59 percent in Western European countries. However, in absolute terms, i.e. comparing the national tenth percentile of the income distribution with the median disposable income in the U.S., the figures are less striking. In the early 1990s, the tenth percentile income in a country relative to the U.S. median income ranges from 42 percent in Italy to 49 percent in Sweden and Belgium (compared with, logically, 36 percent in the U.S.). (See their Fig. 3).

dispersion can be explained by observable characteristics of workers and which share of the wage dispersion is within observably homogenous groups. For the U.S., it has been frequently confirmed that only one third of the wage dispersion can actually be attributed to observable characteristics of workers.[63]

The degree of within-group wage dispersion can be considered in two ways. First, one can look at the distribution of observably homogenous groups (e.g. gender-education-age groups) and take weighted averages of an inequality measure (e.g. the ninetieth to tenth percentile ratio of the within-group distribution). Second, one can look at the distribution of residuals of the estimation of the Mincer-equation (e.g. by looking at a certain percentile ratio of the residual distribution). Wage inequality *within groups* is thus often referred to as *residual wage inequality*. A frequently considered measure of wage dispersion *across groups* is the so-called *college premium*, i.e. the ratio of average wages of employees with a college degree to the average wages of employees with a high-school degree only. Other indicators of wage dispersion across groups encompass the wage differential of non-manual and manual workers and the wage differential of white-collar and blue-collar workers, both usually evolving similarly to the college premium. In the following, we look at trends of both across-group and within-group earnings inequality. Moreover, it is interesting to examine if these trends in earnings dispersion have also affected the *share of labor in national income* and the *average level of wages*, respectively.

Let us begin with *trends in wage dispersion across groups*. In the last three decades, the U.S. and Great Britain share a similar trend concerning the college premium and the non-manual/manual wage differential, respectively.[64] In both countries, wage differentials decreased in the 1970s, but then substantially increased. For example, according to Autor, Katz and Krueger (1998), U.S. college graduates earned 55 percent more than high-school graduates in 1970, saw this premium falling to 41 percent in 1980, before it started to rise up to 65 percent in 1995. Consistent with these findings, Johnson (1997, Tab. 2) reports that the wage of college graduates relative to high-school graduates in the U.S. decreased in the 1970s by an average annual percentage rate of 0.7, but increased by a rate of 1.3 in the 1980s. Relative supply of college graduates not only increased in the 1970s (at an annual rate of 4.8 percent), but also in the 1980s (at an annual rate of 2.7).[65] According to Welch (1999), the share of the population with not more than high-school education aged 55-64 has been 83.2 percent in 1967,

---

63 E.g. Katz and Murphy (1992), Katz, Loveman and Blanchflower (1995).

64 E.g. Katz, Loveman and Blanchflower (1995).

65 Johnson (1997, Tab. 2).

but only 44.9 percent in 1997 among the 45-54 year olds. According to a widely accepted view, since college premiums nevertheless increased in the 1980s, there has been a shift of relative labor demand in favor of college graduates (i.e. high-skilled workers) sufficiently large to outpace the shift in the (relative) supply of skills.[66] In Fig. 1.4, this means that the economy shifts from the initial equilibrium I to the equilibrium III. As a result, both the relative employment level and the relative wage of high-skilled workers increases. In many other countries, consistent with the stable earnings and income dispersion outlined above, the college premium has been fairly stable in the last decades. According to Katz, Loveman and Blanchflower (1995), in Japan and France both the (male) college premium and the non-manual/manual wage differential even fell slightly not only in the 1970s but also until the mid 1980s, just slightly rising afterwards. Edin and Holmlund (1995) establish a similar pattern with respect to the university premium in Sweden, along with a substantial increase in the supply of skills. For Germany, Abraham and Houseman (1995) find that various indicators of male earning ratios between high-, middle and low-education groups have been essentially stable during the 1980s. Also in Germany the skill supply has substantially increased in the last decades. Between 1976 and 1989, the share of the working age population with vocational training has steadily risen from 49.7 to 56.3 percent. Including persons with a university or technical school degree, the share of the working age population with an occupational qualification has even been up from 61 percent in 1976 to 72.2 percent in 1989.[67]

In sum, since an increase in the relative supply of skills in developed countries has not led to falling wage dispersion across groups, one can conclude that relative labor demand has shifted in favor of skilled workers. The question why wage dispersion across groups has increased primarily in the U.S. and Great Britain is still heavily debated. Some authors suggest that the rising U.S. skill premium in the 1980s is due to a decline in the growth rate of skilled labor, whereas wage dispersion has remained stable in countries with a stable or even accelerated skill supply growth.[68] Another hypothesis is that, unlike in the U.S. and Great Britain, strong labor market institutions in Western European countries have prevented wages for low-skilled labor to decline, thus causing unemployment among low-skilled workers.[69] However, there does not seem to be a negative relationship between changes in relative wages and changes in relative employment

---

[66] See Topel (1997) for a detailed discussion of labor supply shifts.
[67] Abraham and Houseman (1995, Tab. 11.10).
[68] E.g. Katz and Murphy (1992), Gottschalk and Smeeding (1997).
[69] E.g. Krugman (1994), Blau and Kahn (1996), Siebert (1997).

levels across countries.[70] Finally, in the U.S. declining union power and falling real minimum wages have been made responsible for the rising wage dispersion. According to Fortin and Lemieux (1997), union membership of male workers in the U.S. declined from 31 percent in 1979 to 21 percent in 1988. Moreover, the U.S. minimum wage, in prices of 1979, has fallen from $2.90 in 1979 to $2.01 in 1989. In fact, Card and Krueger (1995) and Fortin and Lemieux (1997) show that the fall in real minimum wages in the U.S. during the 1980s accounts for over 20 percent of the increase in wage dispersion in that time period. Furthermore, Fortin and Lemieux (1997) find that the decline in union membership has contributed to about 20% of the increase in the variance of (log) wages. Also in Great Britain declining unionizing after the legislative reforms starting in 1979 and rising wage dispersion have occurred at the same time.[71]

Most of the Anglo-American experience in wage dispersion is, however, due to *changes in wage dispersion within groups*. According to Katz and Murphy (1992), Juhn, Murphy and Pierce (1993) and Katz, Loveman and Blanchflower (1995), male within-group wage inequality in the U.S. (measured by the ninetieth to tenth earnings ratio of log weekly wages), increased sharply in the 1970s and 1980s. This leads Katz and Murphy (1992) to conclude that "[...] not only have the less-educated and less-experienced workers lost out [...] but so have the 'least-skilled' or 'least-lucky' workers within each category". Interestingly, unlike the college premium which fell in the 1970s, residual wage inequality started rising from the early 1970s on. In Great Britain, however, within-group wage inequality (measured by gender-age-occupation cells) and the college premium evolved pretty similar, i.e. within-group wage inequality fell in the 1970s and then rose sharply.[72] In Sweden, according to Edin and Holmlund (1995), the explanatory power of the estimation of the Mincer-equation declined from over one third to under one fourth after 1984 which implies that wage dispersion has become increasingly dominant within groups. Moreover, residual wage inequality decreased slightly before that date. This

[70] E.g. Nickell and Bell (1995, 1996), Card, Kramarz and Lemieux (1999). See also Grossmann (1999a) for a discussion of the question whether or not the Anglo-American experience of increasing wage inequality and Western European experience of increasing unemployment are two sides of the same coin. Falkinger and Grossmann (1999) analyze a model in which low-skilled unemployment can occur despite perfectly competitive labor markets and relative employment rates are determined by the average quality of primary education rather than rigidities in the low-skilled labor market. These results suggest that countries can have similar relative employment rates but substantially different relative wages.

[71] See Shackleton (1998) for a discussion.

[72] Katz, Loveman and Blanchflower (1995).

pattern is similar to the evolution of the university premium in Sweden. However, the fact that this is not the case for the U.S. shows that wage inequality within and across groups may be somewhat distinct economic phenomena which require different explanations (see subsection 1.5.2 below for a discussion).

So far, only trends in earnings inequality have been discussed. But it might be interesting to see how patterns of wage dispersion are reflected in both the *share of labor in national income* and the *average real wage level.* As we have seen in section 1.1, both classical and Marxian economists have been concerned with the evolution of the share of labor in national income. Neoclassical growth models predict the functional income distribution to be stable in steady state. As Kaldor (1956, p. 83) writes:

> The famous 'historical constancy' of the share of wages in the national income - and the similarity of these shares in different capitalistic economies, such as the U.S. and the U.K. - was of course an unsuspected feature of capitalism in Ricardo's day.

However, in the last two decades, "capitalistic economies" may have seen a departure from the "historical constancy" of the functional income distribution. Although Walther (1996) finds an essentially stable share of labor since 1970 for the U.S., it seems to fluctuate considerably in Austria and Germany. In the latter two countries, the share of labor increased by almost ten percentage points between 1970 and 1982. Afterwards it fell around five percentage points to about 60 percent in 1991. With respect to Germany, this result is consistent with a study by Poterba (1997) who finds that in continental Europe the non-labor share in business sector output has risen in the 1980s, in contrast to a fairly stable non-labor share in the U.S. The most marked increase of the non-labor share has been taken place in France, from under 30% in the beginning of the 1980s to almost 40% in the beginning of the 1990s. It is also worth noting that these findings are consistent with a substantial increase in the real interest rate in many developed countries from the end of the 1970s until the end of the 1980s.[73] However, different methods of measuring the share of labor substantially alters its measured evolution pattern. For example, it is debatable whether incomes from business owners should be counted as labor or capital income. Similarly, there is no consensus whether or not employers' contributions to social security should be classified as labor income. According to Krueger (1999, Fig. 1), the U.S. share of labor in the post-war period has been highest around 1970. If one includes the employers' contribution to social

[73] OECD (1997a, Tab. 10.10).

security and two thirds of income of owners of unincorporated businesses, the share of labor fluctuates around 75-80 percent after 1970, only slightly decreasing in the early 1990s. However, if these factors are excluded, the share of labor shows a clear downward trend from over 65 percent to under 60 percent. Excluding all income of business owners, the U.S. share of labor has fallen by 5.6 percentage points between 1988 and 1995.

Similar inconsistencies can be found by looking at average real earnings. Using U.S. real earnings data, Abraham, Spletzer and Stewart (1999) report the following. Looking at a data series (from the 'Current Employment Statistics') which excludes tips and infrequent bonuses as well as wages of supervisory workers real hourly (weekly) earnings *fell* by 8.4 (14.1) percent between 1973 and 1997. However, including these factors, real earnings actually *rose* by 3-10 percent in that time period (depending on other measurement differences among wage series). At least, the different evolution of different wage series are consistent with the findings of a rising education premium and declining earning opportunities of low-skilled workers, given that supervisory workers are primarily high-skilled and non-supervisory/production workers primarily low-skilled.

The notion of SBTC suggests that the aggregate production function has changed. Whether or not this has any impact on the functional income distribution cannot be concluded a priori. However, what one would expect is that the share of *low-skilled* labor in national income has declined. In order to test this hypothesis, Krueger (1999) estimates a Mincer-equation in order to approximate the average wage of a hypothetical worker without any schooling. Whereas he finds the share of the so-estimated wages in national income ('raw labor's share', as he calls it) to be pretty stable between 1959 and 1979, equal to 12.2 percent in 1979, it has subsequently declined to a mere 4.9 percent in 1996.[74] This is not only consistent with the rising U.S. college premium and skill upgrading in the last two decades, but also with the finding of dramatically falling real wages of low-skilled workers. Mishel and Bernstein (1994) find that male high-school graduates saw their real hourly wages decline by 20 percent between 1979 and 1993, while wages of men with less than 12 years of schooling fell by even 27 percent. Murphy and Topel (1997) report that real hourly wages fell most in the bottom percentile of the wage distribution between 1969 and 1993. Moreover, absolute wage levels kept decreasing from the early 1970s on for the bottom six (!) percentiles of the wage distribution. According to OECD (1993, Tab. 5.3), whereas real male earnings of the bottom percentile of the earnings distribution in the 1980s declined sharply in the U.S., it showed a slight upward trend in Great Britain, and even substantially increased in Germany.

[74] Krueger (1999, Tab. 2).

1.5.2 Skill-biased technological change: theory and evidence

Thinking about the 'computer revolution', it seems plausible that this recent technological breakthrough is related to the relative labor demand shifts outlined above.[75] However, the popular notion of SBTC, i.e. the hypothesis that new technologies and skills are complements in production, is so crude and simplistic that it raises more questions than it answers. First, one would like to know how the production process is affected by the new technologies in order to get a more differentiated picture about their impact on labor demand. Second, as Thurow (1998) points out, the skill-bias hypothesis is at odds with the frequently confirmed slowdown of *total* factor productivity.[76] A change in total factor productivity is commonly interpreted as indicator of technological change. So why should technological change have a stronger impact nowadays than 20 years ago when it seems to be less rapid today? Third, is it true that "[s]hifts in labour demand from less-educated to the more-educated are [...] probably synonymous with the process of industrialization" [Gregg and Manning (1997, p. 1178)], or that "new technologies are not complementary to skills by nature, but by design" [Acemoglu (1998, p. 1056)]? In fact, computer technology is also used in supermarkets and fast food restaurants, making *low-skilled* workers more productive. Moreover, history is full of technological inventions replacing rather than complementing skills, from the spinning jenny to the assembly line. Thus, it is necessary to shed light into the nebulous notion of SBTC. In the following, theoretical approaches are discussed which are aimed to explain why new technologies seem to increase the earnings dispersion across and within skill groups. Moreover, empirical studies about the relationship between the adoption of new technologies and the demand for skills are reviewed.

Let us begin with *theoretical explanations for SBTC.* Technology-related labor demand shifts in favor of skilled workers are often treated as exogenously given. As indicated in Fig. 1.4, it then depends on the relative labor supply response to a relative labor demand shift whether or not this leads to increasing wage dispersion across education groups.[77] In contrast, Acemoglu (1998) points to an interesting channel linking the observed increasing supply of skilled labor in many countries to deliberate

---

[75] See e.g. Bresnahan (1999), who, unlike most discussions about the impact of computerization on the wage distribution, does not focus on individual gains of computer users, but on changes in the organization of the white-collar workforce in firms. (See below for further discussion).

[76] See Jorgenson and Stiroh (1999) for a recent discussion.

[77] E.g. Gregg and Manning (1997), Heckman, Lochner and Taber (1998).

investments of research firms in technologies that complement skills. In his model, an increase in the supply of skilled labor raises the demand for these technologies (i.e. a 'market size effect') and thus the demand for a skill-complementing technology a successful innovator (who can act as a monopolist) faces. Even though the *short-run response* of rising relative labor supply leads to a decline of the skilled-unskilled wage differential (see Fig. 1.4 with a shift in relative labor supply but not in relative labor demand yielding intersection point IV), the *long-run response* of technological change on the wage differential and thus a relative labor demand shift may even imply higher wage inequality. That is, drawing a curve through intersection points I and III in Fig. 1.4 gives what Acemoglu (1998) calls the "long-run relative demand curve for skilled labor". The important point is that this curve may be upward sloping. Thus, the model can explain why, along with a steady increase in the supply of skills from the 1970s on, the college premium in the U.S. and Great Britain fell in the 1970s (the short-run response) but increased sharply afterwards (the long-run response).

However, more interesting than the explanation of the Anglo-American time path of wage inequality (which, as we have seen below, has been different from that of other countries, raising doubts on the generality of Acemoglu's theory), is the possibility of a 'reversed causality' between an increase in the relative proportion of skilled workers and wage inequality across skill groups. According to a standard view, wage inequality is needed to provide incentives to invest in human capital, and thus is an engine of growth.[78] However, as we have seen above, this view is at odds with the evidence of a negative relationship between inequality and growth. Thus, it may be worthwhile to think about channels suggesting that skill upgrading may rather foster wage inequality.[79]

In their models of endogenous SBTC, Galor and Tsiddon (1997), Caselli (1999) and Lloyd-Ellis (1999) focus on rising skill-requirements, i.e. minimum skills which are necessary for workers to use a certain technology. In the model of Lloyd-Ellis (1999), if the (endogenous) innovation rate exceeds the (exogenous) rate at which workers accumulate skills, wage inequality increases. Moreover, wages increase for workers who fulfill the skill-requirements to work in research. As a result, research costs rise and the innovation rate declines. Thus, the model can resolve the observed paradox that a technology-related rise in wage inequality has taken place at the same time when the total factor productivity has slowed down (see

[78] E.g. Welch (1999).

[79] See also Acemoglu (1999), who argues that an increased availability of skilled workers makes it profitable for firms to create jobs especially designed for skilled workers as the search costs for skilled workers decline.

above). Rather than with ongoing innovations, Galor and Tsiddon (1997) and Caselli (1999) are concerned with major technological breakthroughs, which may be more consistent with the notion that the 'computer revolution' has caused the relative demand for skills to rise. In the model of Galor and Tsiddon (1997), the arrival rate of a new output technology depends on both individual investment in human capital and occupational choices of workers. A technological breakthrough raises wage inequality by increasing skill-requirements. However, if a new technology becomes more accessible over time (e.g. computer software becomes more user friendly), wage inequality declines. In the model of Caselli (1999), workers differ in their ability to acquire the necessary skills to work in new sectors. (In his model, a new sector emerges any time a 'technological revolution' takes place). A new sector is characterized by both higher total factor productivity and a higher capital-labor ratio, whereas the capital-labor ratio decreases in old sectors after a 'technological revolution'. As wages positively depend on the capital-labor ratio, his model not only explains a rising skill premium, but also why wages of low-skilled workers in the U.S. have fallen in absolute terms (see above). Rather than with endogenous technological innovations, Vogt (1999) is concerned with the firms' decision of *adopting* skilled or unskilled labor-saving technologies, respectively. He shows that an exogenous increase in the wage differential of skilled and unskilled labor, e.g. due to declining bargaining power of low-skilled workers, induces firms to raise their skilled labor-saving rate of technological change above the rate of unskilled labor-saving technological change. If the elasticity of substitution between skilled and unskilled labor exceeds unity, this leads to a further increase in the skill premium.[80]

Not only relative wage shifts across observable skill or occupation groups, but also increases in within-group wage inequality have been related to technological change. Heckman, Lochner and Taber (1998) and Caselli (1999) stress that the (unobservable) ability to acquire new skills after a technological shock is not perfectly correlated with observable education levels. Also in Acemoglu (1998, 1999) workers are heterogeneous with

[80] See the appendix of chapter 4 for a formal derivation. Thus, labor market deregulation like declining real minimum wages and declining unionization rates may hurt unskilled workers also because it induces technology adoption to their disadvantage. Note that this is in sharp contrast to the standard view that skill-biased technological progress has caused deteriorating relative positions of low-skilled workers in the first place. The model of Vogt (1999) thus gives a rationale for the fact that in the U.S. and Great Britain rising wage dispersion has been much more dramatic than anywhere else although these countries do not have a better employment record for the low-skilled than other developed countries.

respect to two dimensions, i.e. formal education and actually acquired skill levels (where the latter is unobservable).[81] Lindbeck and Snower (1999) focus on the increasing importance of social competence, communication skills, and the ability to perform multiple tasks. Thus, advances in information and production technologies not only affect relative demand for observable but also for unobservable skills, which may increase residual wage inequality. The maybe most illuminating, although informal, discussion about technology-based shifts of within-group wage inequality can be found in Bresnahan (1999). Bresnahan (1999) argues that the wide-spread adoption of information and communication technologies has raised the demand for highly skilled managers and professionals in "the organisation of bureaucratic production at the firm, industry, and even multi-industry level" (p. F391). Rather than being computer users themselves, the winners from the 'computer revolution' would be workers who have "a deep understanding of one's own organisation and one's customer needs" (pp. F409-F410). For instance, these are white-collar workers who are able to use databases of transactions or operational databases from the accounting and control side of bureaucracies in order to reorganize production and work along new and more productive lines. As the necessary cognitive skills to do so are not captured by standard proxies of human capital (like years of schooling, experience etc.), computerization would drive increases in residual wage inequality.

We have seen that the theoretical literature on SBTC addresses some important stylized labor market facts. But what is the empirical evidence for SBTC? The widely-adopted empirical strategy to confirm that technology-related factors (rather than e.g. trade liberalization) have driven relative labor demand shifts is *(i)* to find evidence for a substantial shift towards high-skilled workers *within* firms and industries,[82] and *(ii)* to show that this

---

[81] By endogenizing skill supply, Acemoglu (1998) is capable to explain why residual wage inequality in the U.S. has increased steadily from the 1970s on, while the college premium fell during the 1970s.

[82] In contrast, a labor reallocation *between* industries would serve as evidence for trade-related factors according to standard trade theory [see e.g. Grossmann (1999a) for a discussion]. To discriminate between employment shifts within and between industries and firms, the total change in the non-production employment share, denoted ΔP_N, is formally decomposed in two terms: $\Delta P_N = \sum_i \Delta S_i \overline{P}_{N_i} + \sum_i \Delta P_{N_i} \overline{S}_i$. S_i denotes the share of total employment in firm or industry i, P_{N_i} the non-production labor share in firm or industry i, and $\overline{S}_i$ and $\overline{P}_{N_i}$ the time average of both variables over the two periods in which the shift in the non-production employment share is considered. The first right-hand-side term

skill-upgrading is positively related to both R&D expenditure and the use of computers and modern information technologies in firms and industries. Due to better data availability, skill-upgrading is a approximated by a shift towards *non-production* employment. Non-production activities encompass coordination and supervision of work (i.e. management), administration, sales, delivery and R&D, among others. There now is overwhelming evidence for both *(i)* and *(ii)* in the developed world.

	Change in non-production employment share		*Change in non-production Wage-bill share*	
	1970-1980	**1980-1990**	**1970-1980**	**1980-1990**
Austria	0.46 (89%)	0.16 (68%)	0.69 (93%)	0.36 (76%)
Australia	0.40 (89%)	0.36 (92%)	0.07 (51%)	0.42 (92%)
Belgium	0.45 (74%)	0.16 (96%)	0.77 (86%)	-0.06 (92%)
Denmark	0.44 (86%)	0.41 (87%)	0.12 (42%)	0.64 (89%)
Finland	0.42 (83%)	0.64 (79%)	0.27 (82%)	0.70 (83%)
Germany	0.48 (93%)	n.a.	0.67 (95%)	n.a.
Great Britain	0.41 (91%)	0.29 (93%)	0.39 (91%)	0.62 (92%)
Japan	n.a.	0.06 (123%)	n.a.	0.14 (84%)
Sweden	0.26 (70%)	0.12 (60%)	0.38 (81%)	0.07 (25%)
U.S.	0.20 (81%)	0.30 (73%)	0.19 (86%)	0.51 (76%)

Table 1.3: Total annualized change (industry level) of the non-production employment and wage-bill share in percentage points and the fraction of the within-industry change in OECD countries. (Within-industry change divided by total change in brackets).

Source: Berman, Bound and Machin (1998, Tab. II and III).

Notes: n.a. = not available. See also footnote 82 for further explanations.

Evidence for (i): According to Machin and van Reenen (1998, Tab. I) both the employment share and the wage-bill share of non-production workers in manufacturing industries steadily increased between 1973 and 1989. For instance, the non-production employment share increased from 24.6 to 30.3 percent in the U.S., from 26 to 32.5 percent in Great Britain, from 25.1 to 31.8 percent in Denmark and from 27.1 to 30.3 percent in Sweden. In the same time period also the non-production wage-bill share increased considerably: from 33.7 to 41.4 percent in the U.S., from 31.7 to 41.4

of this expression denotes changes between, the second one changes within industries or firms. Dividing the second right-hand-side term by the total change ΔP_N gives the shift-share within firms or industries, also reported in Tab. 1.3.

percent in Great Britain, from 33.6 to 40.2 percent in Denmark and from 35.6 to 39.6 percent in Sweden.

Also Tab. 1.3 shows that many OECD countries experienced substantial skill-upgrading in the 1970s as well as in the 1980s, and an overwhelming proportion of this shift has taken place within rather than between industries. Moreover, in the 1980s the move towards non-production employment has taken place despite rising or stable non-production/production wage differentials.[83]

Although most authors concentrate on the broad occupational classification into non-production versus production workers rather than into high-skilled and low-skilled workers, there is evidence that the empirical results indeed reflect skill-upgrading. Machin, Ryan and van Reenen (1996) find an increase in the employment share of high- and medium-skilled workers in the U.S. and Great Britain, mainly within industries, whereas the low-skilled employment share has decreased. Moreover, non-production employment shares in the U.S. and Great Britain are highly positively related to employment shares of high-skilled and medium-skilled workers.

Evidence for (ii): There also is a lot of evidence that the ratio of computer investment to total investment, the fraction of workers using a computer at work and the ratio of R&D expenditure to output are highly correlated with skill-upgrading in manufacturing industries. First, it is interesting to note that, according to Autor, Katz and Krueger (1998, Tab. IV), the share of U.S. workers who directly use computers at work roughly doubled for all skill and occupation groups between 1984 and 1993 (the average share increased from 25.1 to 46.6 percent). Moreover, computer usage is positively related to workers with higher education levels and more prevalent in white-collar than in blue-collar jobs. For instance, 70.2 percent of workers with at least a college degree and 67.6 percent of white-collar workers used a computer in 1993, opposed to 34.6 percent of high school graduates and 17.1 percent of blue-collar workers. Second, although it has been frequently confirmed that skill-upgrading is also positively related to a change in the ratio of *total* investments to output, which hints to a capital-skill complementarity, an increase in the total investment rate can only explain a small fraction of the observed employment shifts towards non-production [e.g. a fraction of 15 percent in the 1980s according to Berman, Bound and Grichilis (1994)].[84] Third, according to Machin, Ryan and van Reenen (1996), the change in the ratio of R&D expenditure to 'value added' can account for 83 percent of the overall change in the Swedish skill

---

[83] Berman, Bound and Machin (1998, Tab. II).

[84] For further evidence that physical and human capital are complementary, see e.g. Bergström and Pannas (1992) and Goldin and Katz (1998).

structure in manufacturing in the 1980s. Equivalent figures for Denmark, the U.S. and Great Britain are 39 percent, 27 percent, and 19 percent, respectively. Interestingly, the impact of technological change on skill-upgrading seems to be lowest in the U.S. and Great Britain, presumably the countries with the least regulated labor markets. This kind of international difference is confirmed by Machin and van Reenen (1998), using both R&D intensity and computer usage. As trade-related factors also seem to have little explanatory power, this finding is rather puzzling. In contrast, using more disaggregated data, Berman, Bound and Grichilis (1994) find that the fraction of investment devoted to computers alone explains over 40 percent of the increase in the non-production wage-bill share in the U.S., whereas computer investments and the ratio of R&D expenditure to sales together account for 70 percent of this shift in the 1979-1987 period. Autor, Katz and Krueger (1998) find a similar explanatory power of this R&D variable together with computer usage for the skill-upgrading in the U.S.

1.5.3 Conclusion and outlook to chapter 4

First, in this section empirical evidence with respect to levels and changes of earnings inequality as well as the (personal and functional) income distribution has been discussed. Whereas rising income inequality and earnings inequality across skill groups has mainly been an Anglo-American experience, residual wage inequality and unemployment of low-skilled workers have risen almost everywhere in the developed world. There is also evidence for a decline in the labor share of national income. Second, it has been pointed out that changes in relative labor demand in favor of better qualified workers are attributed to SBTC in models which focus on shifts in the aggregate production function. Third, it has been argued that empirical evidence suggests substantial skill-upgrading within most industries, and this shift seems to be closely related to computer usage.

However, it remains unsolved whether relative labor demand shifts towards workers who use a computer really reflect the return on computer skills or whether high-skilled workers (with other than special computer skills) just happen to use a computer at work. DiNardo and Pischke (1997) actually find that the wage premium for computer usage is not higher than that of using a calculator, pencils or sitting on a chair at work (!). Clearly, their findings support the second view, showing that computer usage is a quite meaningless indicator for technological change.[85] Thus, wage differentials

[85] DiNardo and Pischke (1997) conclude that "the results seem to suggest that computer users possess unobserved skills which might have little to do with computers but which are rewarded in the labor market, or that computers were first

across skill-groups seem to be associated with white-collar jobs rather than with productivity differentials in the usage of new technologies. Moreover, skill-upgrading within firms and industries is highly related to shifts towards non-production employment (e.g. managers and professionals). This points to a *reorganization of work* within firms and industries rather than to a shift in the production function.[86] On the one hand, the demand for social skills may have risen due to the increased gains from multi-tasking, information sharing and job rotation.[87] On the other hand, the demand for management skills may have risen due to the increased gains from organizational computing like the creation of operational databases.[88] Rather than being computer skills itself, these skills seem to be complementary to computers in organizations. Thus, in order to understand the empirical evidence with respect to both a reallocation of skilled labor towards non-production and the increasing reward for holding a non-production or white-collar job, respectively, a theoretical model at least has to distinguish between production and non-production activities.[89] That is, it has to be taken into account that resources which are not directly related to production activities or the installation of computer equipment, respectively, have to be expended in order to reap the benefits of new information and communication technologies.

---

introduced in higher paying occupations or jobs'' (p. 292). Their results convincingly challenge the view that computer users itself are the winners from computerization, as sometimes has been suggested. [See Krueger (1993) for the seminal paper on this issue, which is supporting this view].

86 See Aghion, Caroli and García-Peñalosa (1999) and Snower (1999a,b) for a discussion about the reorganization of firms and first ideas concerning a possible link to wage inequality.

87 Lindbeck and Snower (1996, 1999).

88 Bresnahan (1999).

89 Non-production activities are neglected in models of SBTC. Moreover, the models of endogenous SBTC reviewed above are primarily designed to explain the U.S. experience: Acemouglu (1998) explains why the U.S. college premium fell in the 1970s and increased in the 1980s. Galor and Tsiddon (1997) argue that the soaring use of electronics in the last two decades have led to increasing skill premiums in the U.S. Lloyd-Ellis (1999) aims to explain why rising wage dispersion occurred simultaneously with the slowdown in aggregate productivity, whereas Caselli (1999) addresses the question why not only the skill premium increased in the U.S. but also why absolute real wages of less-skilled workers declined. However, as discussed above, the U.S. experience of developments in wage inequality has been substantially different than patterns in, say, Western European countries (maybe with the exception of Great Britain). Thus, the models are at odds with the notion that technological change should be similar across developed countries, and thus have similar effects.

Chapter 4 presents a model which simultaneously explains a shift towards non-production work and a decline in the relative position of low-skilled workers with respect to both wages and employment.[90] Although non-production activities are specified as R&D in the model, they may be interpreted to reflect any activity which is undertaken to raise the firms' productivity or revenue, respectively. Shifts towards non-production employment may occur within firms due to an increase in work coordination requirements, care of information networks, marketing, training of workers etc., or across firms due to the outsourcing of business services.[91] Moreover, as chapter 2 has examined the role of status concerns for the income distribution, chapter 4 analyzes how equity concerns affect the distribution of wages. It is shown that the wage structure is determined by social comparisons (i.e. fairness considerations) of workers across as well as within their skill group. Firms have to observe these fairness considerations in setting wages in order to motivate workers to provide high effort. It will be analyzed how social comparisons across and within skill groups interact concerning their consequences for both wage inequality and unemployment of low-skilled labor. It should be noted that macroeconomic consequences of within-group social comparisons have not been considered in the literature yet. In chapter 4, the role of social comparisons is analyzed in a framework in which, as empirically confirmed, high-skilled labor is reallocated towards non-production.

---

[90] See Falkinger (2000) for a new employment theory (in which workers remain jobless despite flexible wages) which is capable of simultaneously explaining a reorganization of work towards decentralized activities and an increase in unemployment.

[91] For a model which relates the relative demand for low-skilled workers towards outsourcing and increased coordination requirements in the context of globalization, see Burda and Dluhosch (1998).

Chapter 2. Status concerns, inequality, and growth

2.1 Motivation

This chapter explores the relationship between *status concerns* and growth. *Social status* is a good which is always in fixed supply, even in a growing economy. Thus, on the one hand, it is interesting to examine how competition among individuals over this scarce good affects people's well-being in the developed world (i.e. in economies where the 'basic needs' are satisfied) as the economy grows. On the other hand, one would like to know how the growth rate of the economy is affected by status-seeking.
The hypothesis that status orientation and social comparisons can substantially affect economic behavior in general and growth in particular has become fairly popular. As Weiss and Fershtman (1998, p. 802) argue:

> Social status is a ranking of individuals (or groups of individuals) in a given society, based on their traits, assets and actions. [...] the interest in social status as a factor in explaining behavior stems from the empirical observation that there is a substantial agreement among different members of society about the relative position of a given individual (or social position). It is this concordance of rankings which gives social status its force as an incentive mechanism. A person of high social status expects to be treated favorably by other individuals with whom he might engage in social and economic interactions. [...] Because of these social rewards, each individual seeks to increase his social status through group affiliation, investments in assets (including human and social capital) and the appropriate choice of actions.

Recently, the growth effects (i.e. the "investments in assets" as Weiss and Fershtman (1998) refer to it) of status-seeking have received a lot of attention (see below for related literature). That is, the impact of individual attempts to gain social esteem (within a certain reference group) on savings and investment behavior and thus on wealth accumulation and growth, respectively, has been extensively explored. The literature on this topic at least goes back to the famous work of Veblen (1899 [reprinted 1934]) who states: "In order to hold the esteem of men it is not sufficient merely to possess wealth or power. The wealth and power must be put in evidence, for esteem is awarded only on evidence." (p. 36). Veblen argues that wealthy individuals consume conspicuous goods and services, thereby signaling their wealth levels in order to gain social esteem. It has frequently been argued that this kind of behavior is similar to a 'rat race' which leads to inefficiently low savings. That is, if everybody tries to improve his/her status by increasing consumption, all may end up with similar *relative* consumption

levels (i.e. similar social status) but lower savings.[1] In turn, lower aggregate savings and thus lower investments ceteris paribus reduce growth.
However, it will be shown in this chapter that savings and growth are not necessarily negatively affected by *relative consumption preferences*. This is because, besides the 'rat race' effect, individuals may be willing to trade current consumption for their future status. Moreover, the impact of status on *voting behavior* with respect to redistribution among heterogeneous individuals (i.e. in an unequal economy) is explored. Redistribution is assumed to take place from owners of (human or physical) capital to owners of a non-accumulated factor such as land or unskilled labor.[2] The latter factor is assumed to be supplied inelastically. The voting outcome potentially affects saving incentives, and thus investment-driven growth. Thus, the political process may be an indirect channel through which relative consumption preferences affect growth. The model builds on the endogenous growth model with majority voting of Bertola (1993), which has already been outlined in chapter 1 (section 1.4). It will be shown that applying this model allows to take a first step towards a positive theory of economic policy in the context of status-seeking and growth [see also Grossmann (1998)].

2.1.1 Some evidence for the presence of status concerns

In this subsection, empirical evidence is presented which can be interpreted to reflect the existence of status concerns and social comparisons. Based on this evidence, it will particularly be argued why the not yet identified politico-economic channel of the status and growth relationship may be a particularly interesting one.
Results from *happiness studies* have usually been interpreted as strong (although indirect) evidence for the presence of individual concerns about relative standing.[3] Happiness studies indicate that, on average, economic

---

[1] This effect has been stressed in the earlier literature about relative consumption preferences. See e.g. Hirsch (1976), Frank (1985a,b).

[2] As will be argued in more detail later, redistribution in the model may best be interpreted as redistribution through progressive taxation of earnings, i.e. from more-educated workers (with high levels of human capital) towards less-skilled/unskilled workers [see Alesina and Rodrik (1994) for a similar interpretation]. In the real world redistribution decreasingly takes place (if at all) from wealth/capital income towards labor income. Taxation of both non-human capital income and wealth is cut almost everywhere in the developed world. In fact, this is challenging for most politico-economic models of redistribution.

[3] E.g. Frank (1985a,b). See Holländer (1996) for further discussion.

growth does not significantly contribute to people's well-being in the developed world.

It should be noted, however, that self-reported happiness *across* countries (rather than *within* countries over time) is strongly positively related to national income. Fig. 2.1 plots the per capita income of 31 countries 1990 and the subjective well-being from a four point scale.[4]

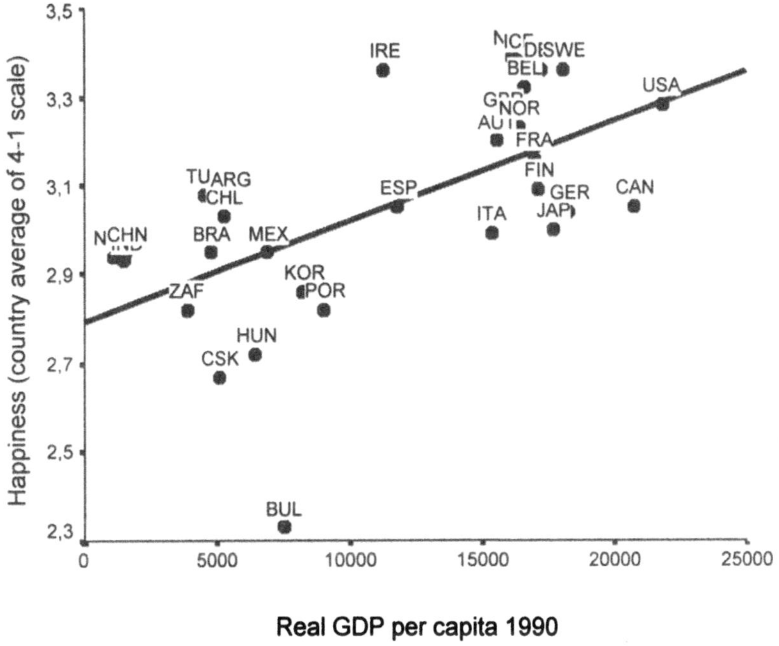

Figure 2.1: Real income per capita and average happiness across countries. *Data sources:* GDP: PWT 5.6. Happiness index: World data base of happiness.

Using OLS and including a constant term, the impact of GDP per capita on average happiness is significant at the 1% level (adjusted $R^2 = 0.337$). Even adding a dummy for (current or former) socialist countries, in which people

[4] Income data shown in Fig. 2.1 are in current international prices and stem from the Penn World Tables 5.6. (An older version of this data set is described in Summers and Heston (1991)). Self-reported happiness observations come from the 'World database of happiness' [Veenhoven (1998)], described in Veenhoven (1995). The question underlying Fig. 2.1 has been: "Taken all together, would you say you are: a) very happy, b) quite happy, c) not very happy, d) not at all happy." [Veenhoven (1998, question 111B)].

are significantly less happy, does not change this result (but considerably improves the fit: adjusted $R^2 = 0.51$).
However, there is plenty of evidence that happiness and satisfaction did not increase in most developed countries *over time*. Fig. 2.2 shows the average score of happiness in the U.S. after the second-world war from a three point scale.[5]

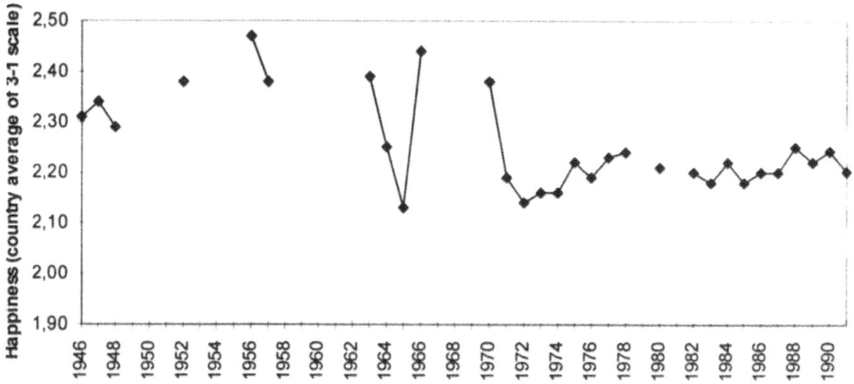

Figure 2.2: Happiness in the U.S. over time.
Data source: World data base of happiness.

Despite enormous economic growth, if anything, there is a downward trend in happiness (although it is remarkable how happy people have been on average in the entire post-war period).[6] Over the whole period (1946-1991), the time trend is negatively related to self-reported happiness (the estimated coefficient is statistically significant at the 1% level, including only time and a constant term in the regression; adjusted $R^2 = 0.306$). However, only looking at the period from 1970 on, the relationship vanishes, showing no trend at all.
As shown in Fig. 2.3,[7] except in Italy, the average level of *personal satisfaction* from a four point scale shows no clear upward trend in other

[5] The question underlying Fig. 2.2 has been: "Taking all things together, how would you say things are these days - would you say you are: a) very happy, b) pretty happy, c) not too happy." [Veenhoven (1998, question 111A)]. If more than one observation is available for a year, simple averages over all observations within this year are taken.
[6] This confirms the findings of Easterlin (1974, 1995).
[7] The question underlying Fig. 2.3 has been: "On the whole, are you: a) fully satisfied, b) not fully but to an extent satisfied, c) as yet unsatisfied, d) very unsatisfied." [Veenhoven (1998, question 121B)]. If more than one observation is

developed countries as well.[8] As could already been seen in Fig. 2.1, there are substantial differences across countries. For example, people in Denmark seem to feel much better than in Japan. Since time series evidence does not reveal that growth makes people happier on average, one can conclude that other factors than income cause cross-country differences in happiness and satisfaction. Maybe the differences merely arise from different perceptions and definitions of these terms across cultures.

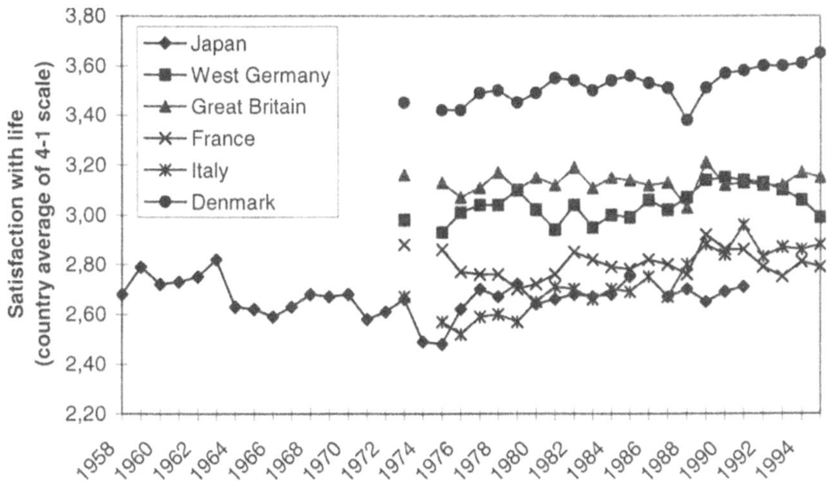

Figure 2.3: Satisfaction with personal life in developed countries over time. *Data source:* World data base of happiness.

What is important about the evolution of subjective well-being in developed countries over time is that it deals with (average) individual 'utility' in the context of growth. In terms of economic modeling, under the assumption that preferences can be represented by an utility function which is identical for all individuals, average happiness may be viewed as an average value of that utility function. If higher material consumption does not make people happier on average, an utility function which is strictly increasing in absolute consumption levels *must* contain other arguments as well, i.e.

---

available for a year, simple averages over all observations within this year are taken.

[8] This confirms a recent finding of Kenny (1999) who regresses subjective well-being against per capita income (rather than against time) for various developed countries. Interestingly, Kenny (1999) also reports evidence that happiness positively affects growth rather than vice versa.

arguments that counter the positive impact of rising consumption on average utility. One candidate of such a countering factor is a negative consumption externality, i.e. individual utility not only increases with the own level of consumption, but also decreases with the consumption of others. To see this, consider the following model which heavily draws on Buchholz (1994). Buchholz (1994) shows that in the presence of a negative consumption externality, growth does not necessarily make people better off, thus formalizing the notion of a 'rat race' among individuals.
There are $n > 1$ identical individuals with an exogenous individual income $y > 0$. There are two goods, a conspicuous consumption good and another good which does not serve as status good. Individual consumption of the status good is assumed to create a negative externality, i.e. it reduces the well-being of other individuals. For simplicity, prices of both goods are exogenous and equal unity. Let c^i denote the conspicuous consumption level of individual *i*. Assume that preferences can be represented by a twice differentiable and concave utility function u which is increasing in the individual consumption of both goods, but decreasing in the total amount of conspicuous consumption of all other individuals. The latter is taken as given by each individual. Each individual *i* simultaneously maximizes

$$u(y - c^i, c^i, \sum_{j \neq i} c^j) \tag{2.1}$$

with respect to $c^i \geq 0$. Note that in a (Cournot-Nash) equilibrium, each individual must consume the same level of the status good, denoted $\tilde{c}$, due to the symmetric situation. Assuming an interior solution, $\tilde{c}$ is implicitly defined by the first order condition:

$$-u_1(y - \tilde{c}, \tilde{c}, (n-1)\tilde{c}) + u_2(y - \tilde{c}, \tilde{c}, (n-1)\tilde{c}) = 0\,. \tag{2.2}$$

(2.2) implies that in equilibrium conspicuous consumption changes with income y (i.e. with economic growth) according to

$$\frac{d\tilde{c}}{dy} = \frac{u_{11} - u_{12}}{u_{11} + u_{22} - 2u_{12} - (n-1)(u_{13} - u_{23})}\,. \tag{2.3}$$

If $u_{11} < 0$, $u_{22} < 0$, $u_{12} \geq 0$, $u_{13} \geq 0$ and $u_{23} \leq 0$, conspicuous consumption unambiguously grows with income. Does growth also unambiguously raise the well-being of individuals? To answer this question, in the following it is considered how utility $u(y-\tilde{c},\tilde{c},(n-1)\tilde{c})$ changes with income:

$$\frac{du}{dy}=u_1\left(1-\frac{d\tilde{c}}{dy}\right)+u_2\frac{d\tilde{c}}{dy}+u_3(n-1)\frac{d\tilde{c}}{dy}=u_1+u_3(n-1)\frac{d\tilde{c}}{dy}, \tag{2.4}$$

where $u_1=u_2$ from (2.2) has been used for the latter equation. Thus, utility grows with income if and only if

$$u_1>-u_3(n-1)\frac{d\tilde{c}}{dy}. \tag{2.5}$$

(Note that $u_3<0$). If the consumption externality is sufficiently strong or if conspicuous consumption increases sufficiently with income, (2.5) does not hold. A negative consumption externality is thus consistent with the stylized fact that people do not become happier with economic growth (i.e. with increases in y).

Clark and Oswald (1996) present evidence that reported job satisfaction of workers negatively depend on the wage level predicted for a worker by estimating a standard Mincer-equation.[9] This wage is interpreted as 'comparison wage rate'. Other evidence for the notion that people make social comparisons come from *experimental studies*. For instance, experimental studies of both the labor market and the private provision of public goods reveal that individuals are not merely self-interested in the sense that they try to maximize their material payoff at any occasion.[10] In fact, they may not do so even in one-shot games. It has been shown that this kind of behavior can be rationalized by interdependent preferences. That is, individuals are also motivated by the final *distribution* of the material payoff.[11] In fact, interdependent preferences require that individuals make

---

[9] The procedure of estimating such a wage rate has been discussed in chapter 1 (section 1.5). According to Clark and Oswald (1996), the actual wage is only positively related to job satisfaction, if this 'comparison wage rate' enters as additional control in the estimated job satisfaction equation.

[10] E.g. Fehr, Kirchsteiger and Riedl (1993), Fehr and Gächter (1999). A survey of these studies is provided by Fehr and Gächter (1998).

[11] E.g. Fehr, Kirchsteiger and Riedl (1998), Fehr and Schmidt (1999). Moreover, experimental subjects reciprocate behavior, i.e. they are kind to kind persons and unkind to unkind persons. However, Grossmann (2000a) shows that reciprocal behavior can be an equilibrium phenomena even in one-shot games with fully rational and merely self-interested individuals.

social comparisons, which is consistent with the presence of concerns for social status. Moreover, an experimental study by Beckman, Formby, Smith, and Zheng (1997) strongly supports the hypothesis that individuals are willing to trade own income in order to improve their relative position.
It remains to be argued that it is reasonable to examine the impact of status-seeking on growth through both its direct impact on savings (i.e. for given policy variables) and the politico-economic channel. Concerning the direct impact on savings one can fall back on the literature pioneered by Veblen (1899). (See the next subsection for further related literature). Analyzing the impact of status concerns on voting over growth-depressing redistribution is especially interesting in light of the stylized fact that people do not become significantly happier on average as the economy grows. On the one hand, redistribution raises relative consumption levels of individuals who are poorer than the average. On the other hand, if individuals have relative consumption preferences, the growth rate of the economy is of less importance for individual well-being than in standard models. Thus, individuals may be willing to accept slower *growth* of their income in order to improve their status. Our analysis suggests that more status-oriented economies tend to have higher degrees of redistribution, even though the growth-reducing effects of taxation are fully taken into account by voters.

2.1.2 Related literature

The literature on the relationship between status-seeking and growth can be classified into two categories. First, like in the model presented below, the utility function of individuals may contain some rank index like relative consumption, relative income or relative wealth. Second, individuals may not be directly concerned with their relative position per se, but act *as if* they would be, in order to acquire *socially provided private goods* which are not allocated through markets. Examples include mating and friendships, honor, esteem, approval and so on. Formally, these social goods rather than relative standing enter the utility function but it turns out that an individual's relative position is instrumental in order to fare well in the social sector.[12]
Belonging to the first category, Fershtman, Murphy and Weiss (1996) assume that the social status of an individual depends on both the average wage and the proportion of skilled workers in the occupation group of this individual. If the demand for status increases with wealth, growth is discouraged by status preferences since workers with high wealth but not necessarily the one's with high ability acquire schooling. In the model of

[12] See Postlewaite (1998) for a discussion of the methodological advantages of this procedure compared with the one of putting a rank index into the utility function.

Duesenberry (1949) individuals with relatively low income levels are myopic to some degree as they neglect future considerations as long as their consumption levels are below their basic needs. These basic needs are assumed to rise with an economy's average income or consumption, respectively. As a result, the individual consumption level is an increasing function of *relative income*. In Konrad (1992), Corneo and Jeanne (1997) and Futagami and Shibata (1998) utility increases with the *relative wealth* level. With identical utility functions of individuals, relative wealth preferences have an unambiguously positive impact on investment-driven growth. (Differences in the growth effects of status preferences, depending on different ways how social status is achieved, are discussed in section 2.5). However, Futagami and Shibata (1998) show that in the case where individuals differ in their relative wealth preferences, stronger status preferences of one group of individuals may adversely affect the steady state growth rate. Harbaugh (1996), Carroll, Overland and Weil (1997) and Rauscher (1997) examine the link between *relative consumption* preferences and growth (as will be done here as well).[13] Harbaugh (1996) explores how savings are affected by relative consumption preferences when there are exogenous shocks in the average growth rate of the economy. In the model below, growth is determined by saving decisions and is thus endogenous. Concerning the *direct* growth effects of relative consumption preferences (i.e. not through the politico-economic channel), Carroll, Overland and Weil (1997) and Rauscher (1997) come closest to the model in this chapter. Whereas in their models individuals are identical in any respect, in the model below there is heterogeneity in factor endowments and, possibly, in preferences. Rather than investigating a *ratio* social comparison model like these contributions, Knell (1999) considers an *additive* social comparison model, i.e. utility is of the Stone-Geary type and the average consumption level of an individual's reference group negatively enters the utility function. In his overlapping generations framework, Knell (1999) is able to distinguish different degrees of status preferences for young and old individuals. If individuals have stronger status preferences when young and if individuals do not socially compare themselves with poorer agents, status-

[13] Assuming relative consumption preferences, also stock market phenomena [e.g. Abel (1990), Campbell and Cochrane (1995)] and time-series features of consumption data [e.g. Carroll and Weil (1994)] have been addressed. Moreover, Hof (1999) analyzes a Ramsey growth model with endogenous labor supply and relative consumption preferences in order to address the question if status concerns lead to inefficient labor supply of representative individuals. Under plausible assumptions, he finds that individuals choose to work too much compared with the social planning optimum.

seeking depresses savings and growth. Moreover, the generally negative impact of inequality on growth in his model (due to imperfect capital markets as outlined in section 1.4) is stronger if individuals are more status-oriented. [Section 2.5 compares the model below with that of Knell (1999)].
Regarding the second category of growth models in which individuals have preferences for socially provided private goods, Cole, Mailath and Postlewaite (1992) were the first to provide microfoundations for status concerns. In their model, individuals are endogenously concerned about their relative wealth in order to do well in the 'marriage market'. As in the models of Konrad (1992) and Corneo and Jeanne (1997), this kind of wealth-seeking may lead to over-accumulation of wealth and thus to excessive growth. Corneo and Jeanne (1998, 1999a) use the approach of Cole, Mailath and Postlewaite (1992) to examine the role of *social segmentation* for the relationship between relative wealth preferences, savings and growth.[14] Corneo and Jeanne (1999b) link initial wealth inequality to growth in the presence of status-seeking and incomplete information about wealth levels, but do not find a clear relationship.
The chapter is organized as follows. Section 2.2 outlines the model. Section 2.3 derives the equilibrium growth path for given policy variables when agents have status concerns. In section 2.4 individual policy preferences for factor income taxation (i.e. redistribution) and political outcomes are derived. Section 2.5 discusses how growth effects of status concerns may change with alternative ways to model status concerns. The last section 2.6 concludes.

[14] In Corneo and Jeanne (1999a) individuals are assumed to be fully informed about wealth levels. Social segmentation is said to be strong if social interaction primarily takes place within relatively homogenous income/wealth groups, i.e. inequality is low in each group individuals compete for status. If social segmentation is strong, the marginal reward of wealth accumulation in the social sector is high. This is because, the lower inequality is, the more a given amount of savings raises the relative wealth level. Thus, wealth accumulation and growth is fostered by segmentation among classes. However, according to Corneo and Jeanne (1998), with incomplete information, individuals have to be engaged in wasteful conspicuous consumption activities which do not directly raise utility but signal wealth [as suggested by Veblen (1899)]. The authors show that more segmentation implies lower aggregate conspicuous consumption levels. In turn, this may reduce growth since wealth first has to be accumulated before it can be wasted in signaling activities.

2.2 The model

2.2.1 Aggregate economy

Consider a closed economy in which a non-decreasing (social) marginal productivity of capital is the source of endogenous growth, as proposed by Romer (1986). There are many identical firms of mass unity. Firms are producing output Y with an accumulated factor K and a non-accumulated factor L. The latter can be viewed as land or unskilled labor whereas the former can be viewed as a composite of human and physical capital.[15] From now on, the accumulated factor will be referred to as 'capital', and the fixed factor will be referred to as 'labor'. The firms' technology at time t is represented by the production function

$$Y(t) = aA(t)K(t)^{1-\alpha}L^{\alpha}\,,\ a>0\,,\ 0<\alpha<1\,, \tag{2.6}$$

where a is a productivity parameter. There is an external productivity of $A(t) = K_A(t)^{\alpha}$, commonly interpreted to be generated by learning-by-doing or human capital spill-over effects. The aggregate capital stock K_A is taken as given by the firms. Since firms have a mass equal to unity, $K_A = K$ holds in equilibrium. The resulting social production function is thus given by $Y(t) = aK(t)L^{\alpha}$. Note that despite increasing social returns to scale, the external productivity in (2.6) allows to maintain the assumption of perfect competition in the goods market since the technology exhibits constant returns to scale for a given level of A. Since savings and thus investments are endogenous in the model, the fact that total productivity rises while capital accumulates may be called *endogenous technological change* or *investment-driven growth*, respectively. The specification of the Romer-type technology in (2.6) is chosen for its familiarity and simplicity.

Following Bertola (1993), the government can redistribute income by imposing linear taxes with tax rates τ_K and τ_L on capital and labor income, respectively. The government budget is assumed to be balanced in any point of time.

Profit maximization implies that the pre-tax returns on capital and labor are equal to $(1-\alpha)Y/K = (1-\alpha)aL^{\alpha}$ and $\alpha Y/L = \alpha aKL^{\alpha-1}$, respectively. Hence, since the balanced budget assumption requires

[15] This is common in one-sector endogenous growth models. [See e.g. Rebelo (1991)]. See section 2.5 and appendix 2.C for a discussion of a decomposition of capital in the model.

$\tau_K[(1-\alpha)aL^{\alpha}]K+\tau_L[\alpha aKL^{\alpha}]L=0$, factor taxes are chosen proportionally according to $\tau_K/\tau_L=-\alpha/(1-\alpha)$. Thus, normalizing total labor supply to unity (i.e. $L=1$) the after-tax returns of capital and labor at time t are given by

$$r(t)=(1-\tau_K)\left.\frac{\partial Y}{\partial K}\right|_{K=K_A,\,L=1}=(1-\tau_K)(1-\alpha)a\equiv(1-\gamma)a, \tag{2.7}$$

$$w(t)=(1-\tau_L)\left.\frac{\partial Y}{\partial L}\right|_{K=K_A,\,L=1}=(1-\tau_L)\alpha aK(t)\equiv\gamma aK(t), \tag{2.8}$$

where $\gamma\equiv(1-\tau_L)\alpha$ is the after-tax income share of labor in national income. Since the return to capital is time-independent, we write $r(t)=r$. Note that a full taxation of labor income implies $\gamma=0$. If $\gamma=0$, the after-tax return to capital is equal to the social return to capital a, according to (2.7). Thus, in this case, there is an appropriate capital income subsidy [i.e. $\tau_K=-\alpha/(1-\alpha)$] which internalizes the positive externality of capital accumulation on the economy's total factor productivity.[16]
Income is said to be redistributed from capital to labor if tax rates are chosen in a way that the after-tax return to capital r is lower than the social return a, i.e. if $\gamma>0$. The higher γ, the higher is the level of redistribution.

Remark: Note that since there is no government spending, one factor income is actually subsidized. (Remember that $\tau_K(1-\alpha)=-\tau_L\alpha$). As this may seem to be an unrealistic feature of the model, one should note that the presence of a subsidy is not essential for any of the results in this chapter. To see this, consider the situation where the tax revenue is used to finance public consumption (i.e. total expenditure for publicly provided goods and services) which is equal to a constant fraction g of total income $Y=aK$. (Remember that $L=1$). Thus, denoting the level of public consumption by G, we have $G=gaK=\tau_K(1-\alpha)aK+\tau_L\alpha aK$ or $\tau_K(1-\alpha)+\tau_L\alpha=g$, respectively, implying $\gamma=(1-\tau_L)\alpha=\tau_K(1-\alpha)+\alpha-g$. Hence, $r=(1-\gamma-g)a$ and $w=\gamma aK$. What matters is that there is a trade-off between the taxation of different factor incomes. In this chapter, this trade-

[16] However, as discussed in subsection 2.5.2, such a policy is generally not socially optimal under the given tax scheme if individuals differ in their capital and labor endowments.

off is indicated by variations in γ, holding g constant (i.e., for simplicity, setting $g = 0$). In contrast, in chapter 3, we focus on redistribution through the public provision of goods and services, i.e. on g.

Without depreciation, capital grows over time according to

$$\dot{K}(t) = aK(t) - C(t), \tag{2.9}$$

where C denotes aggregate consumption. A balanced growth path (which will be derived below) requires

$$\vartheta = \hat{K}(t) = \hat{Y}(t) = \hat{C}(t) = \hat{w}(t) = a - C(0)/K_0, \tag{2.10}$$

where the hat over a variable denotes its growth rate and $K(0) = K_0 > 0$ denotes the initial aggregate capital stock. Note that (2.10) implies that the initial level of aggregate consumption is given by

$$C(0) = (a - \vartheta)K_0. \tag{2.11}$$

2.2.2 Individual budget constraints and preferences

There is a unit-mass continuum of infinitely living consumers, indexed by $i \in [0,1]$, who privately own the production factors. Hence, the aggregate level of a variable denotes its per capita level. Individuals differ in capital endowments $k^i(0) \equiv k_0^i > 0$ and labor endowments $l^i > 0$. Individual budget constraints are given by

$$\dot{k}^i(t) \le rk^i(t) + w(t)l^i - c^i(t) \text{ and } \lim_{t \to \infty} e^{-rt} k^i(t) \ge 0, \tag{2.12}$$

where $c^i(t)$ denotes individual consumption at time t. The latter constraint in (2.12) is the usual 'No Ponzi Game' condition.

Remark: To be more concrete, one may interpret the accumulated factor as human (rather than physical) capital and k_0^i as initial skill (rather than initial wealth) level of an individual. (Note that k_0^i has to be assumed to be strictly positive for technical reasons. However, with the interpretation of k_0^i as innate ability, this seems plausible). According to this interpretation, saved income can be used by individuals to acquire schooling. That is, according to (2.12), one unit of foregone consumption can be transformed in one unit of human capital. As discussed in Romer (1990, p. 78), the labor endowment may be interpreted as "skills such as eyehand coordination", whereas human capital is a "distinct measure of the cumulative effect of activities such as formal education".[17] Not unrealistically, one may think of the case where all individuals hold similar endowments of 'raw' labor, i.e. $l^i = 1$ for all *i*, but have different innate skill levels k_0^i. Thus, the tax system considered here may be interpreted as progressive income taxation which redistributes income from owners of human capital towards owners of unskilled labor.

Preferences are represented by a time-separable utility function

$$(2.13) \qquad U^i = \int_0^\infty e^{-\rho t} u^i\left(c^i(t), c^i(t)/C(t)\right)dt,$$

where ρ denotes the subjective time preference parameter, $0 < \rho < a$. Instantaneous utility u^i may not only be increasing in individual consumption, but also in individual consumption *relative to the per capita consumption level*. Furthermore, marginal instantaneous utility of individual consumption is assumed to be strictly decreasing in c^i for all C. As commonly assumed in the endogenous growth literature, instantaneous utility has a CRRA form, specified as

$$(2.14) \qquad u^i\left(c^i, c^i / C\right) = \begin{cases} \dfrac{\left((c^i)^{1-\beta^i} (c^i / C)^{\beta^i}\right)^{1-\sigma} - 1}{1-\sigma}, \sigma > 0, \sigma \neq 1, \\ (1-\beta^i)\ln c^i + \beta^i \ln(c^i / C), \sigma = 1, \end{cases}$$

where $0 \leq \beta^i < 1$. If $\beta^i > 0$, what matters for an individual is not only the absolute level of consumption but also being above or below the per capita level of consumption. This captures the idea that individuals care about their social status and is referred to as 'status preferences'. Alternatively, one

[17] Distinguishing between the quantity and quality of the labor force as it is done here is also common in growth accounting [e.g. Gollop and Jorgenson (1980)], and, as suggested by Krueger (1999), in measuring the 'raw' labor share in national income (see chapter 1, section 1.5).

could imagine that individuals care about their percentile rank in the distribution of consumption spending. However, it seems more plausible to assume that status is also determined by the actual distance of consumption levels between individuals. The parameter σ denotes the elasticity of marginal instantaneous utility of individual consumption with respect to c^i.

2.3 Equilibrium growth

In this section, the impact of status preferences (i.e. $\beta^i > 0$ for some i) on the equilibrium growth rate ϑ for given tax rates is examined.
Each individual maximizes utility (2.13), (2.14) subject to the budget constraints (2.12), perfectly foreseeing and taking as given the path of aggregate consumption, i.e. each agent perceives himself/herself as too small to have an impact on aggregate consumption.
As shown in appendix 2.A, each individual i chooses his/her consumption level to grow according to

$$\hat{c}^i(t) = \frac{r - \rho + \beta^i(\sigma - 1)\hat{C}(t)}{\sigma}. \tag{2.15}$$

If aggregate consumption grows at a positive rate (i.e. $\hat{C} > 0$), (2.15) implies that individual consumption growth is non-decreasing (non-increasing) in status preferences β^i if $\sigma > 1$ ($\sigma < 1$). This can be interpreted as follows. Consider the case in which the elasticity of the marginal utility of individual consumption is high, i.e. $\sigma > 1$. That is, in the benchmark case $\beta^i = 0$ the intertemporal elasticity of substitution of an individual is low.[18] Because of his/her 'impatience', he/she chooses a low savings rate. Given any aggregate consumption path $C(t)$, this implies a high relative consumption level today at costs of a low one in the future. However, if $\beta^i > 0$, this is not optimal anymore since a small consumption growth rate implies that the individual cannot 'catch-up-with-the-Joneses' in the future. Thus, all other things being equal, he/she chooses a higher savings rate than someone without status concerns. In contrast, if $\sigma < 1$, status preferences induce individuals to give up relative consumption in the future in order to improve status today.
With respect to an equilibrium one can draw the following conclusions:

[18] Note that most empirical studies estimate the intertemporal elasticity of substitution to be lower than unity [e.g. Hall (1988)].

Proposition 2.1 (Steady state growth):
(i) If $\beta^i = \beta$ for all i or $\sigma = 1$, then there exists a unique steady state. The steady state growth rate of the economy is

(2.16) $$\vartheta = \frac{r-\rho}{\sigma(1-\beta)+\beta}.$$

It increases, does not change, decreases with status preferences β if $\sigma >,=,< 1$, respectively.[19]
(ii) If $\beta^i \neq \beta^j$ for some $i \neq j$ and $\sigma \neq 1$, then a steady state does not exist.

Proof:
(i) For $\sigma = 1$ the result directly follows from (2.15) since $\hat{c}^i(t) = r - \rho$ for all i and t. Now consider the case $\sigma \neq 1$ with $\beta^i = \beta$ for all i. Suppose there is an agent i choosing, say, $\hat{c}^i(t) > \hat{C}(t)$ at date t, which is equivalent to $\hat{C}(t) < \dfrac{r-\rho}{\sigma(1-\beta)+\beta}$ according to (2.15). Then there is at least one agent $j \neq i$ choosing $\hat{c}^j(t) < \hat{C}(t)$, which is equivalent to $\hat{C}(t) > \dfrac{r-\rho}{\sigma(1-\beta)+\beta}$. Since this is a contradiction, $\hat{c}^i(t) = \hat{C}(t)$ for all i and t. Substituting the latter expression into (2.15) leads to the result.
(ii) If a steady state exists, then all individuals must choose a common consumption growth rate. According to (2.15), this implies $\vartheta = \dfrac{r-\rho}{\sigma(1-\beta^i)+\beta^i}$. But if $\sigma \neq 1$, this expression depends on i. Hence, it cannot represent a steady state if $\beta^i \neq \beta^j$ for some $i \neq j$ and $\sigma \neq 1$. □

Proposition 2.1 shows that the *direct* effect (i.e. for a given redistribution policy) of status preferences on savings behavior, and thus on growth, is ambiguous. This ambiguity may be explained as follows. On the one hand, status preferences induce individuals to save less due to their desire to catch up with current consumption levels of other people. On the other hand, rational agents know that savings enable them to increase their relative consumption in the future. Stated differently, an individual increase in

[19] Note that there are no transitional dynamics to a steady state growth path with the technology in (2.6).

current consumption has two external (and opposing) effects. First, a *negative consumption externality* (as in section 2.1) which arises since no individual accounts for the negative effect on the *current* status of all others when increasing current consumption, and second, a *positive savings externality* which arises since no individual accounts for the positive effect on the *future* status of others when lowering savings today.
Proposition 2.1 is similar to the results in Carroll, Overland and Weil (1997) and Rauscher (1997), who, however, only consider an equilibrium with *identical* individuals (i.e. $c^i(t) = C(t)$ for all i and t).[20]
Regarding the relationship between inequality and growth, since all individuals choose the same consumption growth rate, the *distribution* of capital and labor endowments, respectively, does not affect the economy's growth rate when policy variables are exogenous.

2.4 Preferred growth rates and redistribution

In this section the impact of both the factor income distribution and status preferences on the voting outcome with respect to redistribution and their consequences for growth are considered. Even if utility functions are such that status preferences do not affect savings decisions in equilibrium for given policy variables (i.e. if $\sigma = 1$), concerns for relative consumption certainly alter preferred redistribution policies. Formally, status preferences enter the indirect utility functions and thus affect desired tax rates.
In order to conduct a useful analysis, only those cases in which a steady state exists are considered. For simplicity, $\beta^i = \beta$ for all i is assumed also in the case $\sigma = 1$. Substituting the interest rate r from (2.7) into (2.16), the steady growth rate of the economy becomes

$$\vartheta = \frac{(1-\gamma)a - \rho}{\sigma(1-\beta) + \beta}. \tag{2.17}$$

As can be seen from (2.17), there is a negative relationship of the labor share γ and the growth rate ϑ, since redistribution from capital to labor income unambiguously depresses the return to capital and thus savings and investment-driven growth.

[20] Carroll, Overland and Weil (1997) even conclude that relative consumption preferences always encourage growth since they misleadingly do not consider the case $\sigma \leq 1$.

To ensure positive initial aggregate consumption $C(0)$ even if labor income is fully taxed (i.e. if $\gamma = 0$), one has to assume[21]

$$(2.18) \qquad a > \frac{a-\rho}{\sigma(1-\beta)+\beta} \equiv \vartheta^{\max}, \text{ i.e. } \rho > (1-\beta)(1-\sigma)a.$$

according to (2.11) and (2.17). Both individual consumption levels and labor income are growing at the same rate ϑ.[22] Furthermore, each individual chooses to accumulate capital at rate $\hat{k}^i(t) = \vartheta$.[23] Hence, using (2.7), (2.8) and the budget constraints (2.12), initial consumption of individual i is given by[24]

$$(2.19) \qquad c^i(0) = \left((1-\gamma)a - \vartheta\right)k_0^i + \gamma a K_0 l^i.$$

(Note that (2.18) implies $r = (1-\gamma)a > \vartheta$, thus ensuring both bounded life-time consumption and bounded life-time labor income). The first term of (2.19) is initial capital income minus savings and the second term is initial labor income. Hence, optimal consumption expenditure exceeds labor income in any point of time.[25] (Remember that $k_0^i > 0$). In fact, with the interpretation of the accumulated factor as human capital or skill level,

---

[21] Remember that $C(0) = (a-\vartheta)K_0$, according to (2.11).

[22] This is because $\hat{c}^i(t) = \hat{K}(t) = \hat{w}(t) = \vartheta$ if a steady state exists, according to (2.10), (2.15) and proposition 2.1.

[23] This can be seen by dividing the first budget constraint in (2.12), which holds with equality, by $k^i(t)$, substituting (2.7) and (2.8), and using $\hat{c}^i(t) = \hat{K}(t) = \vartheta$.

[24] Note that $r - \vartheta > 0$ implies $\lim_{t\to\infty} e^{-rt}k^i(t) = 0$ (transversality condition) since $k^i(t) = k_0^i e^{\vartheta t}$. Integrating the first budget constraint in (2.12), which holds with equality, leads to $\int_0^\infty c^i(t)e^{-rt}dt + \int_0^\infty \dot{k}^i(t)e^{-rt}dt = r\int_0^\infty k^i(t)e^{-rt}dt + \int_0^\infty w(t)l^i e^{-rt}dt$. Using the transversality condition, it is easy to show that $\int_0^\infty \dot{k}^i(t)e^{-rt}dt = r\int_0^\infty k^i(t)e^{-rt}dt - k_0^i$. Substitute the latter expression, (2.7), (2.8), $c^i(t) = c^i(0)e^{\vartheta t}$ and $K(t) = K_0 e^{\vartheta t}$ into the integrated budget constraint to obtain (2.19).

[25] This is a common feature of infinite horizon endogenous growth models. See e.g. Bertola (1993) for a discussion.

respectively (see the discussion in subsection 2.2.2), it is not implausible that individuals spend more than they earn from their 'raw' labor endowment.
According to the balanced budget assumption, the definition of γ and (2.17), the choice of one tax rate (either τ_K or τ_L) determines both the other tax rate, the labor share γ and the growth rate ϑ. Thus, in order to derive the individually preferred redistribution policy, the optimization problem of an individual has to be formulated as depending on one of these variables and exogenous parameters of the model only. In the following, we focus on the growth rate. Rewriting (2.17) as $\gamma a = a - \rho - \vartheta(\sigma(1-\beta)+\beta)$ and substituting this expression in (2.19) yields initial consumption as function of ϑ and the exogenous parameters only:

$$\text{(2.20)} \qquad c^i(0) = \left[\left(\rho - \vartheta(1-\beta)(1-\sigma)\right)k_0^i / K_0 + \left(a - \rho - \vartheta(\sigma(1-\beta)+\beta)\right)l^i\right]K_0 .$$

It will be useful to define the relative factor endowment of individual i:[26]

$$\text{(2.21)} \qquad \xi^i \equiv \frac{k_0^i / K_0}{l^i} .$$

The individually chosen savings rate is given by

$$\text{(2.22)} \qquad sav^i \equiv \frac{\dot{k}^i(t)}{rk^i(t) + w(t)l^i} = \frac{\vartheta\xi^i}{(1-\gamma)a\xi^i + \gamma a},$$

according to (2.7), (2.8) and the fact that $\hat{k}^i(t) = \hat{K}(t) = \vartheta$. Note that sav^i is constant over time and strictly increasing in ξ^i. An individual savings rate coincides with the aggregate propensity to save ϑ / a if and only if $\xi^i = 1$. In this sense, an individual with $\xi^i = 1$ may be called 'representative'.
Social status of agent i is defined by the relative consumption level which is constant over time and given by

[26] Remember that $L = 1$.

$$(2.23) \qquad S^i \equiv \frac{c^i(0)}{C(0)} = l^i \cdot \frac{\left(\rho - \vartheta(1-\beta)(1-\sigma)\right)\xi^i + a - \rho - \vartheta\left(\sigma(1-\beta)+\beta\right)}{a-\vartheta},$$

according to (2.11) and (2.20). S^i is strictly decreasing in ϑ (and thus strictly increasing in γ) if and only if $\xi^i < 1$.[27] (From now on, those individuals are referred to as 'capital-poor'). In other words, lowering the labor income tax rate (and thus depressing the net return to capital due to the trade-off between taxation of capital and labor income) allows capital-poor agents to increase their status. The reason is that relative income, given by $y^i / Y = \left(\xi^i + (1-\xi^i)\gamma\right)l^i$ rises with γ if $\xi^i < 1$. In contrast, if an individual is representative, a change in factor income taxation has no impact on his/her status at all.

Let V^i denote the indirect life-time utility of agent *i*. Regarding (2.13) and substituting the optimal path of individual consumption as well as the aggregate consumption path into (2.14), one obtains

$$(2.24) \qquad V^i + \frac{1}{\rho(1-\sigma)} = \frac{(l^i K_0^{1-\beta})^{1-\sigma}\left[(\rho - \vartheta(1-\beta)(1-\sigma))\xi^i + a - \rho - \vartheta(\sigma(1-\beta)+\beta)\right]^{1-\sigma}}{(1-\sigma)(\rho-\vartheta(1-\beta)(1-\sigma))(a-\vartheta)^{\beta(1-\sigma)}}$$

if $\sigma \neq 1$. [Note that life-time utility is bounded for all i under the assumption in (2.18)]. If $\sigma = 1$, this expression becomes $V^i = \frac{1}{\rho}\left((1-\beta)\ln K_0 + \ln l^i + \frac{(1-\beta)\vartheta}{\rho} + \ln(\rho\xi^i + a - \rho - \vartheta) - \beta \ln(a-\vartheta)\right)$.

It can be shown that V^i is strictly concave in ϑ if $\sigma \leq 1$ or if ($\sigma > 1$ and) $\xi^i < \bar{\xi} \equiv \left(\sigma(1-\beta)+\beta\right) / \left((1-\beta)(\sigma-1)\right)$. Note that, according to (2.20), we also have $\partial c^i(0) / \partial \vartheta < 0$ if $\sigma \leq 1$ or if ($\sigma > 1$ and) $\xi^i < \bar{\xi}$. Thus, if $\sigma \leq 1$ each agent faces a trade-off between his/her initial consumption level and subsequent growth. This means that in choosing the utility maximizing tax policy (to be voted on) individuals have to regard that, on the one hand, higher capital income taxation reduces the rate of consumption *growth*, but,

---

[27] Using (2.23), one finds $\frac{\partial S^i}{\partial \vartheta} = \frac{-[\rho-(1-\beta)(1-\sigma)a](1-\xi^i)l^i}{(a-\vartheta)^2}$. (2.18) implies that the term in square brackets is positive.

on the other hand, it raises the (initial) consumption *level*. However, if $\sigma > 1$, only individuals with a sufficiently small relative factor endowment face such a trade-off. (Note that $\bar{\xi} > 1$ if $\sigma > 1$, i.e. $\xi^i \leq 1$ is sufficient for an individual to face such a trade-off for all σ). If $\sigma > 1$, individuals with a sufficiently large relative factor endowment $\xi^i \geq \bar{\xi}$ unambiguously prefer the lowest possible capital income tax, i.e. the highest possible growth rate $\vartheta^{\max}$. (In this case, V^i is convex).

Regarding the desired growth rates of individuals, two restrictions are plausible. First, we have $\vartheta \leq \vartheta^{\max}$. This follows from the assumption that the preferred tax rate of an individual cannot correspond to a labor income tax rate which exceeds 100%, i.e. $\tau_L \leq 1$ and thus $\gamma \geq 0$, respectively. [Remember the definition of $\vartheta^{\max}$ in (2.18)]. Second, assume that investments are irreversible, i.e. the growth rate cannot be negative. Note that $\vartheta \geq 0$ is equivalent to $\gamma \leq 1 - \rho / a < 1$ (remember that $\rho < a$), according to (2.17), and thus $\tau_K \leq 1 - \dfrac{\rho}{a(1-\alpha)}$, according to the fact that $1 - \gamma = (1-\alpha)(1-\tau_K)$ [see (2.7)], respectively.

Denote the preferred growth rate of individual i under the considered redistribution scheme by ϑ^i, where $\vartheta^i = \arg\max_{\vartheta} V^i$ s.t. $0 \leq \vartheta \leq \vartheta^{\max}$. Furthermore, define $\tilde{\vartheta}^i \equiv \arg\max_{\vartheta} V^i$. Note that $\vartheta^i = \tilde{\vartheta}^i$ if $0 \leq \tilde{\vartheta}^i \leq \vartheta^{\max}$.

We now analyze how the preferred growth rate depends on the relative factor endowment and status preferences of an individual, respectively.

2.4.1 Preferred growth rates and relative factor endowments

The desired redistribution policy and growth rate, respectively, depend on the relative factor endowment in the following way.

Proposition 2.2: For (2.13), (2.14) with $\beta^i = \beta$ for all *i*.

(i) If $\xi^i \geq 1$, then $\vartheta^i = \vartheta^{\max}$, i.e. representative and capital-rich agents prefer the maximal feasible growth rate.

(ii) If $\xi^i < 1$, then $\vartheta^i < \vartheta^{\max}$. If, in addition, $\vartheta^i = \tilde{\vartheta}^i$, then $\partial \vartheta^i / \partial \xi^i > 0$, i.e. the preferred growth rate of a capital-poor agent increases with his/her relative factor endowment.

Proof: Appendix 2.B.[28]

Proposition 2.2 reproduces the result of Bertola (1993) which still holds, if one allows for status preferences. Intuitively, individuals who are labor-abundant relative to their capital endowment vote for a high tax on their less abundant factor.[29] Since policy preferences are single-peaked, the result suggests the following implications of the voting equilibrium. If the political outcome is assumed to be realized by a one-man, one-vote decision and the decisive voter is capital-poor relative to the representative agent, growth is *ceteris paribus* slower in economies with a more unequal distribution of capital.[30]

Alesina and Rodrik (1994) also find that a lower relative factor endowment (as defined here) of the decisive voter slows down growth.[31] In their model, a capital tax is used in order to finance productive public goods and services. The capital tax lowers the return to capital and, if sufficiently large, also the rate of growth. (Small taxes enhance growth since public expenditure is productive). Moreover, capital income taxation raises the *level* of wages. In contrast, in our model there is a trade-off between the return to capital and the *growth rate* of wages. Persson and Tabellini (1994) show that a more unequal income distribution, generated by a more unequal

---

[28] Note that if $\tilde{\vartheta}^i < 0$, then $\vartheta^i = 0$ and ϑ^i may not change with a marginal increase in ξ^i. In general, one can state that ϑ^i is a non-decreasing function of ξ^i, according to proposition 2.2.

[29] Note that, because $\partial V^i / \partial \vartheta = 0$ and $\partial^2 V^i / \partial \vartheta^2 < 0$ in an interior solution, we have $\partial \tilde{\vartheta}^i / \partial \xi^i > 0$ if and only if $\partial^2 V^i / \partial \vartheta \partial \xi^i > 0$, according to the implicit function theorem. That is, proposition 2.2 (ii) is equivalent to the fact that the marginal utility of the growth rate increases with the relative factor endowment of capital-poor individuals.

[30] However, according to (2.21), the distribution of relative factor endowments not only depends on the distribution of capital but also on the distribution of the non-accumulated factor endowment (which could also be land rather than unskilled labor). As Bertola (1993) suggests, in unequal societies land ownership is highly concentrated on the politically decisive class. In this sense, an unequal distribution of the non-accumulated factor endowment also corresponds to an unequal distribution of the relative factor endowment. However, this would mean that an individual with a lower relative factor endowment than the median is decisive. [For a discussion of this kind of political bias, see Bénabou (1996)]. See Persson and Tabellini (1992) for empirical evidence about negative growth effects of high concentration ratios of land ownership in a cross-country study.

[31] See chapter 1 (section 1.4) for a survey on the politico-economic literature on inequality and growth.

distribution of (human) capital endowments, leads to a higher taxation of (human) capital in voting equilibrium. Again, this kind of redistribution is detrimental to growth.[32]

Typically the median voter theorem is applied in these models, i.e. the decisive voter is the one with median relative factor endowment, denoted ξ^m. Whereas in an perfectly egalitarian economy we would have $\xi^i = \xi^m = 1$ for all *i*, the real-world distribution of relative factor endowments is skewed such that $\xi^m < 1$, i.e. the median voter is capital-poor (relative to his/her labor endowment). Thus, we can speak of a more unequal capital income distribution, the higher the measure $1-\xi^m$.[33] However, here we are not particularly interested in the identity of the decisive voter, i.e. we do not consider the impact of a *change* in inequality on growth. Rather, given that the economy is unequal in the sense that the decisive voter is capital-poor, the impact of status concerns on growth through the political process for a given *level* of inequality is considered. This is the issue to which we turn in the following. In doing so, the criticism of the lack of empirical support for a negative impact of a change in inequality on growth through the politico-economic channel does not apply with respect to the impact of status concerns.

2.4.2 Preferred growth rates and status preferences

Consider how redistribution and growth are affected by more marked concerns about relative consumption, i.e. by a (marginally) higher β. According to the first part of proposition 2.2, we have $\vartheta^i = \vartheta^{\max}$ if $\xi^i \geq 1$. Thus, the preferred growth rates of both capital-rich and representative individuals are not affected by status concerns. For reasons discussed below, we restrict the analysis of the impact of β on the preferred growth rate of capital-poor individuals to the case $\sigma = 1$:

---

[32] However, as has been extensively discussed in chapter 1, these results have been strongly criticized on empirical as well as on theoretical grounds.

[33] For instance, suppose that ξ^i is lognormally distributed and let the mean of ξ^i be equal to unity. (If $l^i = L = 1$ for all *i*, this would imply a plausible real-world income distribution). In this case, the difference between the median and the mean of ξ^i increases if and only if the Gini coefficient of its distribution increases [e.g. Aitchison and Brown (1966)]. Under these assumptions, $1-\xi^m$ and the Gini coefficient would be equivalent measures of changes in inequality.

Proposition 2.3: For (2.13), (2.14) with $\beta^i = \beta$ for all *i* and $\sigma = 1$. If $\xi^i < 1$ and $\vartheta^i = \tilde{\vartheta}^i$, then $\frac{\partial \vartheta^i}{\partial \beta} < 0$, i.e. the preferred growth rate of a capital-poor agent decreases with his/her concern for status.

Proof: Appendix 2.B.

The intuition of proposition 2.3 lies in the fact that capital-poor agents gain status with higher levels of redistribution, as shown above [for all σ fulfilling (2.18), i.e. in particular for $\sigma = 1$]. Hence, stronger status preferences induce capital-poor agents to vote for a higher capital income tax, although this discourages savings and thus investment-driven growth.[34] With respect to the voting equilibrium, proposition 2.3 suggests that the more marked the status-orientation of agents is in an economy, the higher is the level of redistribution through growth-depressing taxation, if another redistribution scheme is not feasible.[35]
But what about the cases in which $\sigma \neq 1$? Unlike $\sigma = 1$, there is also a direct effect of β on the growth rate (see proposition 2.1). Moreover, for a given growth rate ϑ, β also affects the initial consumption level, according to (2.20), and thus status of an individual, according to (2.23). Although the analysis is not tractable if $\sigma \neq 1$,[36] the following can be said. For capital-poor agents the direct effects of β on the growth rate ϑ and on initial consumption $c^i(0)$ are opposing for both $\sigma > 1$ and $\sigma < 1$. If $\xi^i < 1$, we

---

[34] Note that, because $\partial V^i / \partial \vartheta = 0$ and $\partial^2 V^i / \partial \vartheta^2 < 0$ in an interior solution, we have $\partial \tilde{\vartheta}^i / \partial \beta < 0$ if and only if $\partial^2 V^i / \partial \vartheta \partial \beta < 0$, according to the implicit function theorem. That is, proposition 2.3 is equivalent to the fact that for capital-poor individuals the marginal utility of the growth rate decreases with the strength of status preferences.

[35] Proposition 2.3 may be compared with the results obtained by Corneo and Grüner (1996) who, however, do not analyze growth effects. The authors consider an environment in which higher economic inequality allows middle-class individuals to reduce competition with poor agents in the social sector [with respect to marriage as in Cole, Mailath and Postlewaite (1992)]. Unlike proposition 2.3, if inequality has an informational value, status-oriented middle-class voters may be willing to provide political support to conservative taxation programs.

[36] If $\sigma \neq 1$, the individually preferred growth rates are only given implicitly, as can be seen in appendix 2.B. Unfortunately, trying to obtain comparative static results in this case is hopeless.

have $\partial c^i(0)/\partial\beta < (>)0$ and $\partial\vartheta/\partial\beta > (<)0$ if $\sigma > (<)1$, according to (2.17) and (2.20), respectively. But, as shown above, capital-poor agents face a trade-off between status and growth for *all* σ fulfilling (2.18). Thus, the relationship between preferred growth and status concerns for capital-poor individuals is likely to be negative, as in the case $\sigma = 1$.

2.4.3 The role of a voter's relative factor endowment for status and preferred growth

How does the impact of status preferences on the preferred redistribution policy and thus on preferred growth vary with the relative factor endowment of a voter?

Proposition 2.4: For (2.13), (2.14) with $\beta^i = \beta$ for all *i* and $\sigma = 1$. If $\xi^i < 1$ and $\vartheta^i = \tilde{\vartheta}^i$, then $\dfrac{\partial^2 \vartheta^i}{\partial\beta\partial\xi^i} > 0$, i.e. the negative impact of status concerns on the preferred growth rate of a capital-poor agent is smaller the higher his/her relative factor endowment.

Proof: Appendix 2.B.

The intuition of proposition 2.4 lies in the fact that the trade-off between status and growth for a capital-poor agent becomes worse the lower the relative factor endowment.[37] If one accepts [following Bertola (1993) and Alesina and Rodrik (1994)] that (i) $\xi^m < 1$, (ii) that the median voter is decisive, and (iii) that higher inequality is associated with a higher value of $1-\xi^m$, then one can conclude the following. First, the negative impact of a rise in status-orientation on growth through redistribution of factor income tends to be larger in more unequal societies. And second, the growth-reducing impact of an increase in inequality tends to be larger in economies populated with more status-oriented agents. However, since the relationship between inequality and growth through the politico-economic channel has not been supported by the data, one may prefer to consider the result in proposition 2.4 with respect to the individually preferred growth rate (or redistribution policy, respectively) only, rather than to draw conclusions for a voting equilibrium.

---

[37] Note that (2.23) implies that $\dfrac{\partial^2 S^i}{\partial\vartheta\partial\xi^i} = \dfrac{[\rho-(1-\beta)(1-\sigma)a]l^i}{(a-\vartheta)^2} > 0$.

2.5 Discussion

As already indicated in section 2.1, the impact of status concerns on growth may crucially depend on the assumption about how status is achieved. Hence, in this section the results derived in the preceding sections are contrasted with the impacts of alternative status preferences. Moreover, welfare effects of the voting outcome are discussed.

2.5.1 Other status concerns

Let us begin with the direct impact of status concerns on growth (rather than the impact through voting decisions) which is ambiguous if status is determined by relative consumption levels, according to proposition 2.1.
First, one could imagine that education or occupation enters the utility function as an indicator for social standing, as in Fershtman, Murphy and Weiss (1996). In their model, the individual schooling decision not only depends on the individual ability to acquire human capital, but also on wealth (i.e. education and thus status is a normal good). As a result, status concerns may lead to a misallocation of human capital investments since workers with high ability but low wealth may be crowded out by rich individuals with low ability. Thus, there may be a negative relationship between status-seeking and growth. As shown in Cole, Mailath and Postlewaite (1992), Konrad (1992) and Corneo and Jeanne (1997), with identical preferences among individuals, status-seeking by social comparisons of owned wealth unambiguously encourages savings and thus investment-driven growth.[38] In these models, individual asset accumulation positively depends on relative wealth concerns. Thus, whether individuals care about their relative standing in form of *relative consumption* or about their *relative wealth* (where the latter could also be reinterpreted as the *relative education level* if the accumulated factor is viewed as human capital) is crucial for the impact of status-seeking on savings behavior. In our model, (human or physical) capital investments only have an indirect effect on status by lowering (relative) consumption today and increasing it in the future.
However, in the class of wealth-seeking models in which utility *directly* depends on relative wealth and individuals have infinite planning horizons, there is a conceptual problem since infinitely living individuals can never

[38] However, Futagami and Shibata (1998) show that the relationship between relative wealth preferences and growth becomes ambiguous if individuals differ in their status concerns.

have an incentive to dissave when old. (It would be interesting to analyze an overlapping generations model in which individuals have both direct preferences for relative wealth and bequest motives). In the other class of models in which wealth-seeking is *endogenously determined* by preferences for non-marketed goods [like marriage as in Cole, Mailath and Postlewaite (1992)], one may not like the idea that wealth accumulation is undertaken in order to 'impress' potential mates. Although there might be anecdotal evidence that the rich have an advantage to 'find attractive mates', it is hard to believe that such motives substantially affect an economy's overall growth rate. One thus may conclude that social comparisons based on consumption rather than wealth levels is a more natural formulation of social status.

Here, we have considered an infinite-horizon model with relative consumption preferences whereas Knell (1999) analyzes an overlapping generation model with Stone-Geary utility functions. If status preferences are equal in all periods (like it is the case here), Knell finds no relationship between status-seeking and growth. This is the same result as in the case of $\sigma = 1$ in proposition 2.1 (in general, proposition 2.1 identifies an ambiguous relationship). The model of Knell (1999) should thus be viewed as complementary to the one here.[39]

Throughout this chapter, capital has been viewed as composite of human and physical capital. One could ask if the results of the previous section still hold if human and physical capital are distinguished. Appendix 2.C outlines a simple example with redistributive taxation of physical and human capital income. The analysis suggests that decomposing the accumulated factor in this type of model does not yield further insights about the impact of status-seeking on growth. The reason is that in steady state equilibrium the *after-tax returns* of all accumulated factors must be *equal*. Each individual is thus indifferent to invest in either type of capital, such that status concerns cannot be a motive to vote for redistributive taxation with respect to different types of capital income. Hence, viewing the accumulated factor as broad capital merely simplifies the analysis above without changing the results. What really matters for the results is the presence of a production factor in fixed supply. In other words, it is the distinction between accumulated and non-accumulated factors rather than the distinction between different accumulated factors that drives the results.

---

[39] Like the direct impact of status-seeking on savings and growth, also voting equilibria with respect to tax policies are likely to depend on the definition of status, although, to my knowledge, there is no further literature on this issue in the context of growth yet.

2.5.2 Welfare effects of taxation

How do status concerns affect efficiency and welfare of the economy through the political process? In the model, taxation of labor income is non-distortionary since there is no labor supply decision of households (i.e. labor is supplied inelastically). In a representative agent version of the model, it would be efficient to set the tax rate on labor income equal to 100%, since this would lead to a private return on capital which is equal to the social return to capital a. Does that mean that a voting equilibrium with less than full taxation of labor income is generally suboptimal from a social point of view? Note that in the model a representative agent exists for any distribution of factor endowments. Due to identical preferences and the fact that individual consumption levels are linear in factor endowments at each date, according to (2.19), any individual endowed with the aggregate levels of capital and labor would choose the economy's consumption path. (Remember that also an individual with $\xi^i = 1$ has been called 'representative' since he/she chooses the economy's aggregate propensity to save. Moreover, such an individual would prefer a tax policy such that the return to capital equals a, according to proposition 2.2 (i)). Nevertheless, a representative agent has welfare significance only if lump sum redistribution of endowments is possible. However, the unavailability of lump sum redistribution was the reason to consider the 'distortionary' redistribution scheme in the first place. In fact, any voting outcome in the model is *constrained efficient* in the sense that a social planner could not improve efficiency with this redistribution instrument in hand. Thus, although there is an equality-growth trade-off in the model, there is no equality-efficiency trade-off. Whether or not the voting outcome is *socially optimal* thus crucially depends on the social welfare function one wishes to employ.[40]

2.6 Summary

This chapter has explored the impact of people's concern about their *relative consumption level* on both savings behavior and voting behavior with respect to tax policies in a simple general equilibrium framework with endogenous growth and majority voting.

If agents behave fully rational, individual saving rates were shown to be ambiguously affected by status preferences in an equilibrium with steady

---

[40] For a further discussion, see Putterman, Roemer, Silvestre (1998), Grossmann (1999b).

investment-driven growth, depending on the elasticity of marginal utility of the individual consumption level. However, allowing for endogenous redistribution/tax policy, each agent has a motive to vote for a policy that improves his/her social status. If the decisive (i.e. median) voter is capital-poor relative to the representative agent of the economy, redistribution from capital to labor income rises with the degree of status-orientation of the decisive voter, although that reduces the growth rate of wages unambiguously. This is because in the presence of status-seeking and growth-reducing redistribution, capital-poor individuals face a trade-off between their status and overall economic growth.

Appendix to chapter 2

Appendix 2.A. Derivation of individual consumption growth (2.15)

The current-value Hamiltonian function for the utility maximization problem of an individual denoted i, given his/her initial capital endowment $k_0^i > 0$, is

$$\text{(2.A.1)} \qquad \Im(c^i, k^i, \lambda^i) = \frac{\left(c^i / C^{\beta^i}\right)^{1-\sigma} - 1}{1-\sigma} + \lambda^i (wl^i + rk^i - c^i),$$

where λ^i is the current-value shadow price of individual income of the non-accumulated factor. The first-order conditions

$$\text{(2.A.2)} \qquad \frac{\partial \Im}{\partial c^i} = 0 \Leftrightarrow \lambda^i = \frac{(c^i)^{-\sigma}}{C^{(1-\sigma)\beta^i}},$$

$$\text{(2.A.3)} \qquad -\frac{\partial \Im}{\partial k^i} = \dot{\lambda}^i - \rho\lambda^i \Leftrightarrow -\hat{\lambda}^i = r - \rho$$

and the transversality condition $\lim_{t\to\infty} e^{-rt} k^i(t) = 0$ [where the latter holds because of $r > \vartheta$, according to (2.18)] are necessary and sufficient for a maximum because of the concavity of $\Im$ (for all C) and positive discounting (i.e. $\rho > 0$). Differentiating (2.A.2) with respect to time yields

$$\text{(2.A.4)} \qquad -\hat{\lambda}^i = \sigma \hat{c}^i + (1-\sigma)\beta^i \hat{C}.$$

Combining (2.A.3) and (2.A.4) gives equation (2.15). □

Appendix 2.B. Proof of propositions 2.2-2.4

Proof of proposition 2.2: Using the first-order condition for a maximum of the indirect utility function V^i in (2.24) with respect to ϑ and neglecting the restrictions $0 \le \vartheta \le \vartheta^{\max}$ for a moment, $\tilde{\vartheta}^i = \arg\max_{\vartheta} V^i$ is implicitly defined by

(2.B.1)

$$F(\cdot) \equiv -\left((1-\beta)(1-\sigma)\xi^i + \sigma(1-\beta)+\beta\right)\left(\rho-(1-\beta)(1-\sigma)\tilde{\vartheta}^i\right)(a-\tilde{\vartheta}^i)+$$
$$\left[(1-\beta)(a-\tilde{\vartheta}^i)+\beta\left(\rho-(1-\beta)(1-\sigma)\tilde{\vartheta}^i\right)\right]\times$$
$$\left[\left(\rho-(1-\beta)(1-\sigma)\tilde{\vartheta}^i\right)\xi^i + a-\rho-\tilde{\vartheta}^i(\sigma(1-\beta)+\beta)\right]=0.$$

This expression has the quadratic form $F(\cdot)=x(\tilde{\vartheta}^i)^2+y\tilde{\vartheta}^i+z=0$, where $x=(\sigma-1)(\bar{\xi}-\xi^i)(1-\beta)^2(\sigma(1-\beta)+\beta)$ with $\bar{\xi}$ as given in section 2.4. If the second-order condition for a maximum holds, then $x>0$. [Remember that if $\sigma>1$ and $\xi^i \geq \bar{\xi}$, V^i is convex and $\vartheta^i = \vartheta^{\max}$ (corner solution)]. If $\beta>0$, one finds two roots for $\tilde{\vartheta}^i$ in (2.B.1). To see which one is the relevant root, consider the benchmark case $\beta=0$. According to (2.B.1), if $\beta=0$, then $\tilde{\vartheta}^i = \dfrac{\rho(\xi^i-1)+\vartheta^{\max}}{(1-\xi^i)\sigma+\xi^i}$ [with $\vartheta^{\max}=(a-\rho)/\sigma$, $x=\sigma(\sigma(1-\xi^i)+\xi^i)$ and $y=-x(a+\tilde{\vartheta}^i)$]. Note that $y<0$ if and only if $x>0$. This implies $2\tilde{\vartheta}^i x+y=-x(a-\tilde{\vartheta}^i)<0$. Hence, dealing with real roots of (2.B.1), only the smaller root can be a solution if $\beta>0$, implying

(2.B.2) $$\frac{\partial F(\cdot)}{\partial \tilde{\vartheta}^i}=2\tilde{\vartheta}^i x+y<0 \text{ for all } \beta \geq 0.$$

Furthermore, one obtains

(2.B.3) $$\frac{\partial F(\cdot)}{\partial \xi^i}=\left(\rho-(1-\beta)(1-\sigma)\tilde{\vartheta}^i\right)\times \left[\sigma(1-\beta)(a-\tilde{\vartheta}^i)+\beta\left(\rho-(1-\beta)(1-\sigma)\tilde{\vartheta}^i\right)\right]$$

which is strictly positive according to (2.18). Hence, applying the implicit function theorem, one finds $\dfrac{\partial\tilde{\vartheta}^i}{\partial\xi^i}>0$, using (2.B.2) and (2.B.3). Finally, confirming from (2.B.1) that $\tilde{\vartheta}^i=(a-\rho)/(\sigma(1-\sigma)+\beta)=\vartheta^{\max}$ if $\xi^i=1$ concludes the proof of proposition 2.2. □

Proof of propositions 2.3 and 2.4: If $\sigma = 1$, then (2.B.1) becomes

$$F(\cdot) \equiv (1-\beta)(a-\tilde{\vartheta}^i)^2 + (1-\beta)\rho(\xi^i - 2)(a-\tilde{\vartheta}^i) + \beta\rho^2(\xi^i - 1) = 0. \tag{2.B.4}$$

Thus, one gets an explicit solution for $\tilde{\vartheta}^i$ given by

$$\tilde{\vartheta}^i = a - \rho + \frac{\rho}{2}\left(\xi^i - \sqrt{(2-\xi^i)^2 + \frac{4\beta(1-\xi^i)}{1-\beta}}\right) \tag{2.B.5}$$

(which is again the smaller root). Proposition 2.3 directly follows from (2.B.5). Proposition 2.4 follows from using (2.B.5) to calculate $\partial^2 \tilde{\vartheta}^i / \partial\beta\partial\xi^i$ and making use of $\xi^i < 1$. □

Appendix 2.C. Redistribution between different accumulated factor incomes

Consider the case of factor income redistribution with two different accumulated production factors but without a non-accumulated factor. Let the technology be described by the production function

$$Y(t) = bK(t)^{1-\beta} H(t)^{\beta},\ b > 0,\ 0 < \beta < 1, \tag{2.C.1}$$

where K denotes aggregate physical capital and H denotes human capital. For simplicity, assume that the economy is initially in its steady state equilibrium. [See Barro and Sala-i-Martin (1995, ch. 5) for a discussion of the transitional dynamics in a similar type of model]. Because the after-tax returns of both accumulated factors must be equal in steady state, a balanced government budget implies that $K(t)/H(t) = (1-\delta)/\delta$ for all $t \geq 0$, where δ denotes the after-tax share of human capital income. Thus, both types of capital grow at the same rate. The steady state rate of return to capital $r^* = \delta^{\beta}(1-\delta)^{1-\beta}$ is maximized at $\delta = \beta$ (i.e. both factor income taxes are equal to zero). Let each individual have an initial endowment k_0^i and h_0^i of physical and human capital, respectively, with $k_0^i + h_0^i > 0$. Maximizing individual utility (2.13), (2.14) subject to the individual budget constraints $c^i(t) + \dot{k}^i(t) + \dot{h}^i(t) = r^*\left(k^i(t) + h^i(t)\right)$ and $\lim_{t\to\infty} e^{-rt}\left(k^i(t) + h^i(t)\right) \geq 0$

yields, in case of $\beta^i = \beta$ for all $i \in [0,1]$, a balanced growth rate $\vartheta^* = (1/\sigma)(r^* - \rho)$ [see proposition 2.1]. From individual budget constraints together with the assumption $r^* > \vartheta^*$ it follows that status

(2.C.2) $$S^i = \frac{k_0^i + h_0^i}{K(0) + H(0)}$$

does not depend on ϑ^*. Hence, status concerns cannot be a motive to vote for taxation with respect to different *accumulated* factors.

Chapter 3. Inequality, voting over public consumption, and growth

3.1 Motivation

As has been extensively discussed in the preceding chapters, growth models linking the income distribution and the size of the government are based on majority voting. According to these models, inequality is positively related to the size of the government (due to the demand for redistribution policies) and taxation depresses growth. Whether or not taxation empirically has a negative and significant effect on growth is controversial. Whereas Easterly and Rebelo (1993), Perotti (1996) and Figini (1999) find ambiguous results, Kneller, Bleaney and Gemmell (1999) suggest that when 'productive' public expenditure like infrastructure or education expenditure is controlled for, the ratio of the total tax revenue to GDP negatively affects growth. (See chapter 1, section 1.4 for further evidence and discussion). However, the notion that a negative relationship between income inequality and growth works through the politico-economic channel is not at all confirmed by the data [e.g. Perotti (1996), Figini (1999)].[1] Also without considering growth, the empirical evidence for the hypothesis that a more unequal income distribution leads to higher redistributive government spending [see e.g. Meltzer and Richard (1981) for a theoretical model] is, at best, mixed.[2]
In this chapter, it is suggested that standard majority voting models lack conclusive empirical support because they are based on a too narrow view of redistribution measures.[3] In the real world, the median income earner rarely receives cash or in-kind transfers, as these transfers are mainly provided to

---

[1] For further empirical evidence and discussion, see the recent paper by Basett, Burkett and Putterman (1999).

[2] Whereas Meltzer and Richard (1983) provide some time-series evidence for the U.S. in favor of their theory, Lybeck (1986) rejects the hypothesis for Sweden. In cross-country studies, Mueller and Murrell (1986) find some weak support, but Kristov, Lindert and McClelland (1992) even find that a lower median to mean income ratio *reduces* redistribution.

[3] Other caveats concerning a simple politico-economic relationship between inequality and redistribution have been expressed. For example, Roemer (1998) has shown that when voting problems are multi-dimensional, i.e. there is voting not only over redistribution but also over a non-economic issue, it may well be that poor individuals vote for conservative taxation programs. Moreover, Bénabou (1996) and Lee and Roemer (1998) show in growth models with imperfect capital markets, that redistribution is a U-shaped function of inequality, if the pivotal voter is not the median voter.

poorer individuals. (Think about welfare spending). The relevant redistribution to median income earners may rather be effectuated through taxes levied in order to finance publicly provided goods and services. Examples include recreational facilities, parks, roads, health and cultural services. Tax payments usually rise with income, but the median income earner does not necessarily consume less of these goods and services than richer individuals. As Boadway and Wildasin (1986, p. 506) state: "Almost any taxing or expenditure decision of [...] governments will have distributive implications; it cannot be avoided."

Consequently, this chapter examines the link between income inequality and the size of government in majority voting equilibrium under the consideration of tax-financed redistribution through publicly provided goods and services rather than transfer policies.[4] More specifically, it is analyzed if the share of government spending of national income rises or falls with income inequality through the political process. This is done in both a simple static model and a dynamic general equilibrium model with investment-driven growth. Whereas in the static version individual incomes are exogenous, the intertemporal model also allows to analyze the role of the tax effects on savings for the voting outcome. The dynamic model is similar to the one in chapter 2, except that taxation now finances publicly provided goods and services rather than transfer spending. Thus, the voting outcome derived in this chapter is directly comparable with standard politico-economic growth models like the one in chapter 2.[5]

It should be noted, however, that the majority voting approach in this chapter is not chosen because it is supposed to be the best way to represent actual political processes, but in order to show that even within this framework standard results about the relationship between inequality, growth and the size of government can be overturned.

Like in chapter 2, saving decisions determine the rate of investment-driven growth and agents differ in (human or physical) capital endowments as well as in the endowment of a non-accumulated factor (like unskilled labor or land). As it turns out, there is generally not a positive relationship between income inequality and the public consumption share of national income in

---

[4] Opposed to the long debate about whether or not a higher *per capita income* level yields a larger public consumption share of total income, a hypothesis known as Wagner's law, it is considered how the *income dispersion* affects the size of the public sector.

[5] Remember that the framework in chapter 2 has been chosen because it has been appropriate to analyze the impact of status concerns of heterogeneous individuals on growth rather than the impact of a change in inequality on growth through the political process. Status concerns are not considered in the present chapter.

majority voting equilibrium. Thus, given the hypothesis that the public provision of goods and services rather than cash transfers is the relevant redistribution for median income earners, the model is consistent with a weak empirical link between inequality and tax rates.

Interestingly, due to the different saving decisions of heterogeneous individuals, the impact of inequality on public consumption is rather different in the intertemporal model than in the static one. For instance, if individuals have Cobb-Douglas preferences for private and public consumption and there is a proportional income tax, in the static case the government spending fraction in voting equilibrium is *not* affected by the income distribution. However, with the same assumptions about both preferences and taxation as in the static model, higher inequality of (human) capital income unambiguously *lowers* the public consumption share in the growth model. This is because in a steady state, income from both the accumulated and the non-accumulated factor grow at the same rate, whereas the propensity to save is higher for (capital-)richer individuals. It turns out that in this (not implausible) scenario, a (say) synthetic income tax in particular adversely affects private consumption *levels* of (capital-)poor individuals, whereas the (steady state) income growth rate is affected similarly among individuals. As a result, relatively (capital-)poor individuals demand a rather *low* level of taxation and thus a small public sector. In this sense, the result of the static model that higher inequality does not necessarily yield higher tax rates in voting equilibrium is even strengthened in the growth model.

Furthermore, some rough empirical cross-section estimates are presented, showing that income inequality and the size of public consumption is even negatively related within the OECD. Using a broader set of countries, no significant relationship is found.

In a related paper, Fiaschi (1999) also addresses the empirical failure of politico-economic growth models. His model is based on Alesina and Rodrik (1994) in which government spending is productive. That is, in his model public investment enters the production function of a representative firm which produces a homogenous private consumption good, whereas in our model the level of public consumption enters the utility function of agents. Moreover, rather than showing that inequality need not be positively related to public expenditure as in our growth model, Fiaschi (1999) argues that taxation need not be negatively related to growth in a majority voting framework with heterogeneous agents.[6]

---

[6] However, his suggestion that higher labor income taxation raises growth hinges on the same mechanism as embodied in the models of Bertola (1993) and Alesina and Rodrik (1994) in which tax rates on capital and labor income are negatively

Section 3.2 sets up a simple static median voter model and examines the link between the income distribution and the government spending fraction. Section 3.3 proposes a growth model which is in some sense comparable with the static model of section 3.2. The equilibrium growth rate is derived in section 3.4. Section 3.5 identifies the additional effects on the individual demand for public consumption compared with the static case, whereas section 3.6 derives the voting equilibrium. In section 3.7, the implications of the analysis and the contribution to both the public choice and the growth literature are discussed. Section 3.8 gives some cross-section estimates about the relationship of income inequality and the government consumption share. The last section concludes.

3.2 Voting over public expenditure: a simple model

First, a static model with majority voting over the level of public consumption expenditure is presented. There are $i \in [0,1]$ individuals with an exogenous income $y^i \in (0, y^{\max}]$. Thus, total income $Y = \int_0^1 y^i di$ equals per capita income. Denote c^i the individual consumption level of a private numeraire bundle and G the level of a publicly provided good.[7] (In the following, we speak of a private and a public consumption good, respectively).[8]

---

related. (See also chapter 2 for such a model). In such a framework, an increase in the labor tax rate raises growth since it goes along with a reduction in the capital tax rate. The only difference of the model of Fiaschi (1999) to these standard models is that he derives a negative relation of individually desired tax rates on labor and capital income. In contrast, in e.g. Bertola (1993) a negative relationship between taxation of labor and capital income is directly implied by the redistribution scheme with direct transfers and a balanced government budget. Concerning the impact of the relative factor endowment of the median voter (i.e. inequality) on taxation, the model of Fiaschi (1999) produces exactly the same results as the models of Bertola (1993) and Alesina and Rodrik (1994).

[7] See Epple and Romano (1996) for a different median voter model with a private numeraire bundle and a publicly provided private good. Whereas their paper allows for private supplements of the publicly provided good, our focus is the link between income distribution and the public sector.

[8] The publicly provided good can either be viewed as pure public good or as private good which is consumed in equal amounts by the individuals. The reason for this is the following. If the publicly provided good is private and $G^i = G$ for

Individuals have identical preferences represented by a twice differentiable utility function $u(c^i, G)$ which is increasing in its arguments and strictly concave. Assume that one unit of the private consumption good can be transformed in one unit of the public consumption good. The public consumption good is financed by income taxes. Let $s^i = s(y^i)$ denote the tax share (or contribution share, respectively) to public consumption of individual i, where $s(\cdot)$ is a non-decreasing function. Furthermore, denote the *average* tax rate as a function $\tau(\cdot)$ of income, i.e. $\tau(y^i) \in [0,1]$ is the average income tax rate of individual i. Thus, a balanced budget requires $G = \int_0^1 \tau(y^i) y^i \, di$, and, according to the definition of the tax share, we have

$$(3.1) \qquad s^i = s(y^i) = \frac{\tau(y^i) y^i}{G}.$$ [9]

That is, for a given income y^i, the average tax rate of individual i endogenously adjusts to the level of public consumption G, i.e. $\tau(y^i) = s(y^i)G / y^i$, according to (3.1).

Each individual i solves the following problem, taking his/her tax share s^i as given.

$$(3.2) \qquad \max_{c^i, G} u(c^i, G) \ \text{ s.t. } \ c^i + s^i G \le y^i .$$

Assuming an interior solution, the individual demand for G, denoted G^i, is given by

$$(3.3) \qquad \frac{u_2(y^i - s(y^i)G^i, G^i)}{u_1(y^i - s(y^i)G^i, G^i)} = s(y^i).$$

(3.3) simply states that the marginal rate of substitution between the two goods equals the individual tax share (i.e. the individual 'price' of the public

---

all i, where G^i denotes the individual consumption level of that good, then we have $G = \int_0^1 G^i \, di$ with a unit mass of individuals.

[9] Note that $\int_0^1 s^i \, di = 1$.

consumption good in units of the private consumption good).[10] This condition implicitly defines the individual demand G^i for public consumption as function of individual income y^i. Let individuals vote over the level of publicly consumption *G*. Since policy preferences over the level of public consumption are single-peaked due to the concavity of the utility function, the median voter theorem can be applied. The voting outcome is assumed to be a one-man, one-vote decision. How does G^i depend on the individual income y^i? If there would be a monotone relationship for all y^i, the decisive median voter would be the median income earner. Following the politico-economic literature about income inequality and redistribution, an income distribution is viewed as more equal, the higher the median income is relative to the mean (see chapter 2).
Now consider homothetic preferences, i.e. without loss of generality, assume that the utility function *u* is homogenous.[11] Under this specification, (3.3) can be written as

$$\left[MRS\left(\frac{c^i}{G^i}\right)=\right]MRS\left(\frac{y^i-s(y^i)G^i}{G^i}\right)=s(y^i)\,, \tag{3.4}$$

where $MRS(\cdot)\equiv u_2(\cdot,1)/u_1(\cdot,1)$ denotes the marginal rate of substitution.[12]

---

[10] See e.g. Borcherding and Deacon (1972), Bergstrom and Goodman (1973).

[11] In our context, homothetic preferences can be justified as follows. Assuming both a constant price elasticity and a constant income elasticity of public consumption demand, denoted δ and κ, respectively, we have $G^i=h(s^i)^\delta(y^i)^\kappa, h>0$ [e.g. Bergstrom and Goodman (1973)]. Under the assumption that the decisive voter is the one with median income, i.e. the public consumption good is non-inferior, this expression yields a typical estimated equation in the public choice literature on public consumption, given by $\ln G=\ln h+\delta\ln s^m+\kappa\ln y^m+\eta$, where s^m and y^m is the median tax share and median income of the economy, respectively (η is an error term). Typically, κ (i.e. the elasticity of public consumption with respect to the median income holding the tax share constant) is estimated around unity. [For survey articles about this kind of empirical estimates, see Mueller (1989, ch. 17), Holsey and Borcherding (1997)]. This empirical result is consistent with homothetic preferences.

[12] Note that if $u(c^i,G)$ is homogenous of degree *m*, we have $u_1=G^{m-1}\zeta'(c^i/G)$ and $u_2=G^{m-1}\left(m\zeta(c^i/G)-(c^i/G)\cdot\zeta'(c^i/G)\right)$, where $\zeta(\cdot)\equiv u(\cdot,1)$.

Lemma 3.1: With homothetic preferences, the elasticity of public consumption demand with respect to individual income (regarding that the individual tax share may vary with individual income) is given by

$$\text{(3.5)} \qquad \frac{dG^i}{dy^i}\frac{y^i}{G^i} = 1 - \chi^i \varepsilon^i,$$

where $\varepsilon^i \equiv \frac{d(c^i / G^i)}{d\,MRS(c^i / G^i)} \frac{MRS(c^i / G^i)}{(c^i / G^i)}$ is the elasticity of substitution evaluated at G^i and $\chi^i \equiv s'(y^i)y^i / s^i$ is the income elasticity of the individual tax share.

Proof: See appendix. □

The first term of the right-hand-side of (3.5) is the income elasticity of individual public consumption demand if the individual price of public consumption s^i is held constant.[13] (As it is well known, this elasticity is equal to unity with homothetic preferences). According to the second term of (3.5), if the individual tax share rises with income (i.e. if $\chi^i > 0$), the sign of the *total* income elasticity of public consumption demand is ambiguous (note that $\chi^i \varepsilon^i >,=,< 1$). This is because richer individuals have to pay a higher 'price' for the publicly provided good. Nevertheless, rich individuals may vote for a large public sector. Lemma 3.1 also implies that even with homothetic preferences the relationship between public consumption demand and individual income is not necessarily monotone such that the median income earner is generally not identical with the median voter. In this case, the voting equilibrium depends on the entire income distribution.[14] One can conclude the following.

Proposition 3.1 (Static model): If the individually desired public consumption level G^i is a monotone function of $y^i \in (0, y^{\max}]$, implying that the median income earner is the decisive voter, higher income

---

[13] If, e.g., the public consumption good would be financed by a uniform poll tax, this would be the only effect. Thus, in this case demand for public consumption would be a positive function of individual income [Atkinson and Stiglitz (1980, ch. 10)].

[14] E.g. Bergstrom and Goodman (1973).

inequality (i.e. a lower median income for a given mean income) does not necessarily lead to a larger size of public consumption in voting equilibrium.

Proof: Directly follows from (3.5) and the fact that $\chi^i \varepsilon^i$ may be lower than unity and may depend in any form on y^i. □

Since public expenditure has a redistributive impact if the tax share is an increasing function of income, proposition 3.1 questions the standard argument that redistribution measures and thus tax rates are higher in more unequal economies.

Example (Proportional income tax): In the remainder of the theoretical part of this chapter, proportional income taxes are considered. In the static model above, this means that tax shares s^i are such that $\tau(y^i) = \tau$ holds for all $y^i \in (0, y^{\max}]$. Thus, because $G = \tau Y$ with a constant average tax rate, one obtains $s^i = y^i / Y$, according to (3.1), i.e. the individual tax share equals relative income. This implies $\chi^i = 1$, i.e. the income elasticity of the individual tax share equals unity. Also note that τ equals the public consumption share of total income. Hence, voting over the level of public consumption is equivalent to voting over the public consumption share. Using $s(y^i) = y^i / Y$ and $G = \tau Y$, (3.4) becomes

$$(3.6) \qquad MRS\left(\frac{(1-\tau^i)\theta^i}{\tau^i}\right) = \theta^i,$$

where $\theta^i = y^i / Y$ denotes relative income and $\tau^i = G^i / Y$ denotes the preferred public consumption share of individual i.[15]

Proposition 3.2 (Static model with a proportional income tax): Suppose homothetic preferences and a proportional income tax. If the median income earner m is the decisive voter and $\varepsilon^m < (=,>) 1$, higher income inequality,

[15] Thus, for a given relative income of the median voter, the public consumption share in the politico-economic equilibrium does not depend on the stage of development of an economy. In a median voter model, this is just an alternative way of saying that Wagner's law does not hold.

i.e. a lower median to mean income ratio θ^m, leads to a larger (the same, a lower) public consumption share of total income τ^m in voting equilibrium.

Proof: Using $\theta^i = y^i / Y$ and $\tau^i = G^i / Y$, we have $\frac{d\tau^i}{d\theta^i}\frac{\theta^i}{\tau^i} = \frac{dG^i}{dy^i}\frac{y^i}{G^i}$.

Thus, $\chi^m = 1$ implies $\frac{d\tau^m}{d\theta^m}\frac{\theta^m}{\tau^m} = 1 - \varepsilon^m$ according to (3.5). □

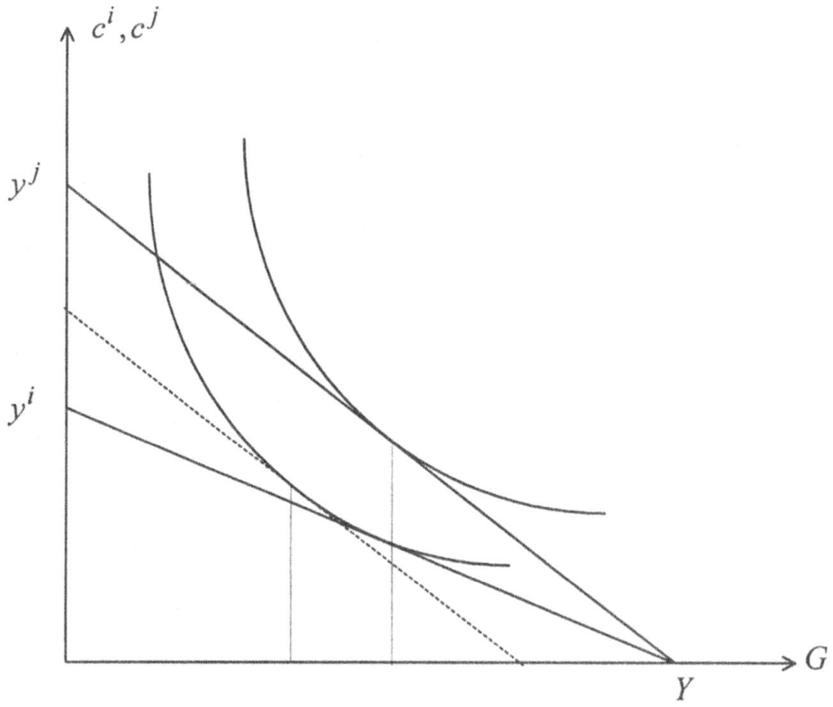

Figure 3.1: Individual demand for a publicly provided good with a proportional income tax and Cobb-Douglas preferences (i.e. $\varepsilon^i = 1$ for all *i*).

The total effect of a higher (relative) income on the desired level of public consumption can be divided into two effects: first, a *substitution effect* arising from an increase in the relative 'price' of the public consumption good faced by an individual (since the individual tax share rises with income), and second, a *wealth effect* (or income effect, respectively) due to the rising consumption possibilities (for a given tax share). Since the

publicly provided good is normal, both effect go in opposite directions. If the individually preferred ratio of private and public consumption is sufficiently elastic (inelastic) to the individual tax share, i.e. if $\varepsilon^i > 1$ ($\varepsilon^i < 1$), the substitution effect dominates (is weaker than) the wealth effect. Fig. 3.1 illustrates proposition 3.2 in the *c*-*G* space for the Cobb-Douglas case (i.e. $\varepsilon^i = 1$ for all *i*) in which all individuals vote for the same level of public consumption. Two individuals *i* and *j* with different incomes are considered, and $y^i < y^j$. As can be seen from Fig. 3.1, the income effect and the substitution effect of a higher tax share exactly cancel in the Cobb-Douglas case.
In the next section, it is examined how this result changes due to different saving decisions of heterogeneous agents in a dynamic general equilibrium growth model in which public consumption is financed by proportional taxes on both accumulated and non-accumulated factor income.

3.3 Public expenditure in a growth model

3.3.1 Aggregate economy

There is a private consumption good *Y* produced by many identical firms and a publicly provided good *G* which is financed by proportional *factor income taxes*. As in chapter 2, output *Y* is produced with an accumulated factor *K* (called 'capital') and a non-accumulated factor *L* (called 'labor'). Again, the accumulated factor may best be interpreted as human capital and the economy's total supply of labor is normalized to unity (i.e. $L = 1$). Output *Y* at time *t* is produced according to

$$(3.7) \qquad Y(t) = aA(t)K(t)^{1-\alpha} L^{\alpha}\,,\ a > 0\,,\ 0 < \alpha < 1\,,$$

where a is a productivity parameter. As in chapter 2, there is an external productivity of $A(t) = K_A(t)^{\alpha}$ and the aggregate capital stock K_A is taken as given by the firms. With a unit mass of identical firms $K_A = K$ holds in equilibrium which implies $Y(t) = aK(t)L^{\alpha}$ for the social production function.
In order to finance the public consumption good the government imposes taxes on labor and capital income with tax rates $0 \le \tau_L \le 1$ and $0 \le \tau_K \le \tau_K^{\max}$, respectively (see below for the definition of $\tau_K^{\max}$). Then at

time t after-tax returns $r(t)$ and $w(t)$ on K and L, respectively, are given by

(3.8) $$r(t) = (1-\tau_K)\frac{\partial Y}{\partial K}\bigg|_{K=K_A,\,L=1} = (1-\tau_K)(1-\alpha)a \equiv r,$$

(3.9) $$w(t) = (1-\tau_L)\frac{\partial Y}{\partial L}\bigg|_{K=K_A,\,L=1} = (1-\tau_L)\alpha a K(t),$$

according to (3.7). The government budget is assumed to be balanced in any point of time. As in the static model of section 3.2, let the marginal rate of transformation between the private and public consumption be constant and normalized to unity. Thus, the quantity of the publicly provided good equals the tax revenue. Using (3.8) and (3.9), we have

(3.10) $$G(t) = gaK(t), \text{ where}$$

(3.11) $$g = \frac{G(t)}{Y(t)} = \tau_K(1-\alpha) + \tau_L \alpha .$$

(Note that $Y(t) = aK(t)$ since $L = 1$). Thus, the government spending fraction of total output, g, is a weighted average of the tax rates. If $\upsilon = 1$ we have the special case of a synthetic income tax. In order to analyze a more general case, consider the following tax scheme

(3.12) $$\tau_L = \tau \text{ and } \tau_K = \upsilon\tau,$$

where the parameter $\upsilon \geq 0$ denotes the constant factor tax ratio (if $0 < \tau \leq 1$).[16] Using (3.11), (3.12) implies the following proportional relationship between the government share of total income g and the tax rate on labor income τ:

(3.13) $$\tau = \frac{g}{\Phi}, \text{ where } \Phi \equiv (1-\alpha)\upsilon + \alpha .$$

Without depreciation, capital grows over time according to

[16] Note that $\tau_L = 0$ implies $\upsilon \to \infty$ for a given $\tau_K > 0$.

(3.14) $$\dot{K}(t) = aK(t) - C(t) - G(t),$$

where *C(t)* denotes the aggregate level of private consumption at time *t*. Thus, if there a balanced growth rate ϑ (which will be derived below), we obtain

(3.15) $$\vartheta = \hat{K}(t) = \hat{Y}(t) = \hat{C}(t) = \hat{w}(t) = \hat{G}(t),$$

where the hat over a variable denotes its growth rate. Moreover, according to (3.10), (3.14) and (3.15), the initial level of aggregate private consumption is given by

(3.16) $$C(0) = \big(a(1-g) - \vartheta\big)K_0,$$

where $K(0) = K_0 > 0$ denotes the aggregate initial capital stock of the economy.

3.3.2 Individual budget constraints and preferences

There is a unit-mass continuum of infinitely living consumers indexed $i \in [0,1]$, who privately own the production factors. Hence, the aggregate level of a variable is equal to its per capita level. Individuals differ in (human) capital endowments $k^i(0) \equiv k_0^i > 0$ (i.e. the individual innate skill levels) and labor endowments $l^i > 0$. Denoting the individual consumption of the private consumption level at time t as $c^i(t)$, the individual budget constraints are given by

(3.17) $$\dot{k}^i(t) \le rk^i(t) + w(t)l^i - c^i(t) \text{ and } \lim_{t\to\infty} e^{-rt}k^i(t) \ge 0$$

(as in chapter 2). Each individual has the following time-separable utility function

(3.18) $$U^i = \int_0^{\infty} e^{-\rho t} u\big(c^i(t), G(t)\big)dt,$$

where ρ denotes the subjective time preference parameter. For technical reasons, assume $0 < \rho < (1-\alpha)a$. As in chapter 2, instantaneous utility has a CRRA form, which here is specified as

$$u(c^i, G) = \frac{\left(c^i G^\gamma\right)^{1-\sigma} - 1}{1-\sigma}. \tag{3.19}$$

Note that (3.19) is a monotone and positive transformation of a Cobb-Douglas (and thus a homogenous) utility function. (Hence, preferences are homothetic). The parameter $\gamma > 0$ indicates the individual preference for public consumption, and $\sigma > 0$ is the elasticity of marginal instantaneous utility with respect to private consumption. As it will turn out, the assumption $\Omega \equiv \sigma(\gamma+1) - \gamma > 0$ ensures that the balanced growth rate is non-negative if $r \geq \rho$, i.e. $\tau_K \leq 1 - \frac{\rho}{(1-\alpha)a} \equiv \tau_K^{max}$, according to (3.8).[17] Two properties of (3.19) are notable. First, the marginal instantaneous utility with respect to private consumption increases (does not change, decreases) with the level of public consumption G, if $\sigma < 1$ ($\sigma = 1$, $\sigma > 1$). And second, (3.19) implies that the elasticity of substitution between the two goods is equal to one, i.e. $\varepsilon^i = 1$ for all i. Thus, due to the latter property of (3.19) and the assumption of the proportional income tax, the growth model is directly comparable with the example of the static model illustrated in Fig. 3.1.

3.4 Equilibrium growth and public consumption

In this section the equilibrium growth rate ϑ for given tax rates is derived. Each individual maximizes utility (3.18) and (3.19) subject to the budget constraints (3.17), perfectly foreseeing and taking as given the path of public consumption, determined by (3.10) and (3.11).

Lemma 3.2: Each individual i chooses the private consumption level to grow according to

$$\hat{c}^i(t) = \hat{C}(t) = \frac{(1-\tau_K)(1-\alpha)a - \rho + \gamma(1-\sigma)\hat{G}(t)}{\sigma}. \tag{3.20}$$

---

[17] Note that $\tau_K^{max} \in (0,1)$, since $0 < \rho < (1-\alpha)a$ has been assumed.

Proof: See appendix. □

Thus, using (3.15), the balanced growth rate is given by

$$\vartheta = \frac{(1-\tau_K)(1-\alpha)a-\rho}{\sigma(\gamma+1)-\gamma} = \frac{(1-\upsilon\tau)(1-\alpha)a-\rho}{\Omega}. \tag{3.21}$$

[Remember $\tau_K = \upsilon\tau$, according to (3.12), and $\Omega \equiv \sigma(\gamma+1)-\gamma > 0$]. First, note that due to the simple technology in (3.7), there are no transitional dynamics to the steady state growth path. Second, as already pointed out in the previous chapters, capital income taxation in this type of model discourages growth since it reduces the private return on investment, in turn reducing the amount of savings. Moreover, the growth rate ϑ does not depend on the taxation of labor income. The reason for this is the following. According to (3.20), all agents desire the same private consumption growth rate which is equal to the growth rate of labor income ϑ in steady state, according to (3.9), (3.15) and (3.21). Since agents are infinitely living, this also implies that each individual accumulates capital at the same rate ϑ.[18] Thus, at each instant, the aggregate amount of savings does not depend on income from the non-accumulated factor.[19] Third, one finds that a higher public consumption preference parameter γ raises (does not affect, lowers) ϑ if $\sigma < 1$ ($\sigma = 1$, $\sigma > 1$). This result can be understood as follows. For instance, if $\sigma < 1$, the marginal instantaneous utility of private consumption increases with current public consumption. Thus, if $\sigma < 1$, it is optimal for an individual to choose a higher growth rate of private consumption (i.e. a higher savings rate) if $\hat{G} > 0$, the stronger the preference for public consumption. [See (3.20)]. The intuition for the cases $\sigma = 1$ and $\sigma > 1$ is completely analogous.

3.5 Individual demand for public consumption and growth

In this section, it is considered how the individual demand for public consumption depends on factor incomes. First, the analysis is carried out for

---

[18] This can be seen by dividing the first budget constraint in (3.17), which holds with equality, by $k^i(t)$, substituting (3.8) and (3.9), and using $\hat{c}^i(t) = \hat{K}(t) = \vartheta$.

[19] See chapter 1 (section 1.4) for a discussion of the critical assumptions for this result.

the case $\sigma = 1$. Second, after having derived the basic mechanisms for $\sigma = 1$, the results are compared with the case $\sigma \neq 1$.[20]
Using (3.8), (3.9), (3.17) and presuming $r > \vartheta$ to obtain bounded life time consumption, initial consumption of individual i is given by $c^i(0) = \left((1-\tau_K)(1-\alpha)a - \vartheta\right)k_0^i + (1-\tau_L)\alpha a K_0 l^i$. (The derivation of the private individual consumption level is completely analogous to chapter 2). Substituting (3.21) we have

$$c^i(0) = \frac{\Gamma(1-\tau_K)+\rho}{\Omega}k_0^i + (1-\tau_L)\alpha a K_0 l^i, \tag{3.22}$$

where $\Gamma \equiv (\sigma-1)(\gamma+1)(1-\alpha)a$. For $\sigma = 1$ this reduces to

$$c^i(0) = \rho k_0^i + (1-\tau)\alpha a K_0 l^i, \tag{3.23}$$

using $\tau_L = \tau$ from (3.12). The first right-hand-side term of (3.22) and (3.23), respectively, is initial capital income minus savings and the second term is initial labor income (net of taxes). Thus, if $\tau < 1$, the optimal private consumption expenditure exceeds labor income in any point of time.[21] (For a discussion about the plausibility of this feature of infinite horizon growth models, see chapter 2). This can be understood by the fact that the capital growth rate is identical among individuals which implies that capital income minus savings does not depend on the labor endowment. If $\sigma = 1$, capital income minus savings does not depend on the private return of investment as well. Thus, in this case only the tax rate τ on the non-accumulated factor matters for the private consumption level. As in chapter 2, it will be useful to define relative factor endowments

$$\xi^i \equiv \frac{k_0^i / K_0}{l^i}. \tag{3.24}$$

[20] This procedure has two advantages over the analysis of both cases, i.e. $\sigma = 1$ and $\sigma \neq 1$, at once. First, the mechanisms driving the results come out more clearly. Second, algebraically the case $\sigma \neq 1$ has to be treated somewhat differently to the case $\sigma = 1$.

[21] Remember $k_0^i > 0$.

(Remember that total supply of labor is normalized to unity). Using this definition, one finds that the individual savings rate[22]

(3.25) $$sav^i = \frac{\vartheta \xi^i}{r\xi^i + \alpha a(1-\tau)}$$

is increasing in ξ^i if $\tau < 1$. Again, this is an implication of the identical capital accumulation rates chosen by individuals.

In order to derive the individually preferred public consumption shares, one has to observe the restrictions on the public consumption share implied by the restrictions on the tax rates. First, since $0 \le \tau_L = \tau \le 1$, we have $0 \le g \le \Phi = (1-\alpha)\upsilon + \alpha$, according to (3.13). Second, under the assumption that investments are irreversible (as in chapter 2) and since there is no capital depreciation, the growth rate has to be non-negative which puts a restriction $\tau_K = \upsilon\tau \le 1 - \frac{\rho}{(1-\alpha)a} = \tau_K^{\max}$ on capital income taxation, according to (3.21). According to (3.13), this implies $g \le \Phi\tau_K^{\max}/\upsilon$. Note that $g \le \Phi\tau_K^{\max}/\upsilon$ (or $g \le \Phi$) can be neglected if $\tau_K^{\max} < \upsilon$ (or $\tau_K^{\max} \ge \upsilon$, respectively).

3.5.1 The case $\sigma = 1$

If $\sigma = 1$, substituting the optimal path of individual private consumption and the public consumption path into (3.18) and using (3.19), we obtain $V^i = (1/\rho)\left(\ln c^i(0) + \gamma \ln G(0) + (1+\gamma)\vartheta/\rho\right)$ for the indirect life-time utility of agent *i*.[23] Substituting (3.10), (3.21) and (3.23) into this expression and using (3.12), (3.13) and (3.24) yields

(3.26)

$$\rho V^i = \ln\left(\rho\xi^i + (1 - g/\Phi)\alpha a\right) + \gamma \ln g + \frac{(1+\gamma)\left((1-\upsilon g/\Phi)(1-\alpha)a - \rho\right)}{\rho} + \Lambda$$

---

[22] The individual savings rate is given by $sav^i = \dot{k}^i(t)/y^i(t)$ with $y^i(t) = rk^i(t) + w(t)l^i$. Using (3.9) and the fact that $\dot{k}^i(t) = \vartheta k^i(t)$ yields (3.25).

[23] Note that if σ approaches unity, instantaneous utility (3.19) becomes $u(c^i, G) = \ln c^i + \gamma \ln G$.

where $\Lambda \equiv \ln(aK_0^2 l^i)$ is an unessential constant. Denote the preferred public consumption share of individual i by g^i, where $g^i = \arg\max_g V^i$ s.t. $g \geq 0$, $g \leq \Phi$, (or) $g \leq \Phi \tau_K^{\max} / \upsilon$. It can be shown that V^i is strictly concave in g.[24] Furthermore, define $\tilde{g}^i \equiv \arg\max_g V^i$. Note that policy preferences with respect to the government spending fraction g are given by $g^i = \tilde{g}^i$ if $\tilde{g}^i \geq 0$, $\tilde{g}^i \leq \Phi$, (or) $\tilde{g}^i \leq \Phi \tau_K^{\max} / \upsilon$.

Lemma 3.3: For $\sigma = 1$. If $g^i = \tilde{g}^i$, the preferred public consumption share g^i is an increasing function of the relative factor endowment ξ^i.

Proof: See appendix. □

The intuition of lemma 3.3 can be seen from the first three terms of (3.26). The first and the second term reflect the impact of the spending fraction g on utility from the initial private and public consumption level, respectively, whereas the third term reflects the impact of general growth on utility. According to the second and third term of (3.26), all individuals face the same trade-off between the initial public consumption level and growth. This is because all individuals choose the same growth rate of private consumption [see (3.20)]. In other words, since individual income from both factors, i.e. capital and labor, are growing at the same rate, the *growth* impact of taxation affects all individuals equally. In contrast, the optimal choice of the initial private consumption *level* depends on the individual factor endowment. First, according to (3.23), only the taxation of labor income affects the private consumption level. (According to (3.22), if $\sigma \neq 1$, the impact of capital taxation on private consumption is ambiguous; see below for the implications). Having established the role of taxation effects on the utility from private consumption, lemma 3.3 is now easy to understand. The less an individual has to rely on labor income, i.e. the lower his/her endowment l^i, the weaker the trade-off between the private consumption level and the public consumption share of aggregate income. Moreover, a higher relative capital endowment k_0^i / K_0 simply reduces the marginal utility of private consumption. In sum, the individual trade-off between the private consumption level and the public consumption share

---

[24] This ensures that the median voter theorem can be applied. (See the next section for the derivation of the voting equilibrium).

decreases with the relative factor endowment. (Formally, this can be seen by substituting (3.13) into (3.23) and confirming that $\partial \ln c^i(0)/\partial g$ is an increasing function of ξ^i).[25]

If public consumption would be financed by capital income taxation only (i.e. $\tau_L = 0$ and $0 < \tau_K \le \tau_K^{\max}$), the private consumption level would not be affected, according to (3.23). As a result, all individuals would vote for the same public consumption share. (It is easy to show that $\tilde{g}^i = (1-\alpha)\tau_K = \dfrac{\rho\gamma}{a(1+\gamma)}$ if $\tau_L = 0$). Remember that for the CRRA utility specification (3.19), the elasticity of substitution between the two goods equals unity. Thus, for $\sigma = 1$ and $\tau_L = 0$, one obtains a similar result as proposition 3.2 (i.e. as in the static model of section 3.2 with a proportional income tax), if $\varepsilon^i = 1$ for all *i*. However, if labor income is also taxed, the result is considerably different.

3.5.2 The case $\sigma \neq 1$

According to (3.22), if $\sigma \neq 1$, initial private consumption is also affected by capital income taxation which potentially alters the result.

According to (3.18) and (3.19), life-time utility becomes $V^i = (1-\sigma)^{-1}\left[\left(c^i(0)G(0)^\gamma\right)^{1-\sigma}/\left(\rho-(1-\sigma)(\gamma+1)\vartheta\right)-1/\rho\right]$.[26] It is easy to check that maximization of V^i is equivalent to maximization of $\ln\left((1-\sigma)V^i + 1/\rho\right)/(1-\sigma) =$ $\ln c^i(0) + \gamma \ln G(0) - \ln\left(\rho-(1-\sigma)(1+\gamma)\vartheta\right)/(1-\sigma)$ with respect to *g*. Thus, as in the case $\sigma = 1$, it suffices to look at the impact of a marginal increase of the public consumption share on the utility of the private consumption level, i.e. on the partial derivative $\partial \ln c^i(0)/\partial g$. This is because, similar to

[25] Note that, because $\partial V^i / \partial g = 0$ and $\partial^2 V^i / \partial g^2 < 0$ in an interior solution, we have $\partial \tilde{g}^i / \partial \xi^i > 0$ if and only if $\partial^2 V^i / \partial g \partial \xi^i > 0$, according to the implicit function theorem. That is, lemma 3.3 is equivalent to the fact that the marginal utility of the public consumption share increases with the relative factor endowment.

[26] Note that according to (3.21), $r > \vartheta$ is equivalent to $\rho > (1-\sigma)(\gamma+1)r$, thus implying that $\rho-(1-\sigma)(\gamma+1)\vartheta > 0$.

the case $\sigma = 1$, the trade-off between the public consumption level and growth is not individual specific.

Lemma 3.4: For bounded life-time utility, i.e. $r > \vartheta$. If $g^i = \tilde{g}^i$, *(i)* the preferred public consumption share of total income g^i increases with (does not depend on, decreases with) the relative factor endowment ξ^i if $\sigma > 1$ and $\upsilon < (=, >) 1 + \rho / \Gamma$, and *(ii)* g^i unambiguously increases with ξ^i if $\sigma < 1$.

Proof: See appendix. □

First, consider the case $\sigma > 1$, i.e. the intertemporal elasticity of substitution with respect to private consumption is lower than unity. In other words, individuals are fairly impatient and therefore choose low savings. Thus, compared to $\sigma = 1$, there is now the additional effect that a higher public consumption share *lowers* initial private consumption through capital income taxation (if $\upsilon > 0$). Obviously, this effect becomes larger the higher the individual capital endowment. As this counters the effect that a higher capital endowment reduces the marginal utility of private consumption, the latter effect dominates only if the factor tax ratio υ is not too high. For instance, the result that the individually preferred public consumption share nevertheless increases with the individual capital endowment holds for the important special case of a synthetic income tax (i.e. $\upsilon = 1$). If $\sigma < 1$, the private consumption level *increases* with the capital income tax rate.[27] This is because $\sigma < 1$ means that individuals are fairly patient and thus reduce their savings by a lower amount than the amount by which capital income declines due to an increase in the capital income tax rate. As a result, the positive impact of a higher (relative) capital endowment on the preferred public consumption share is strengthened compared to the case $\sigma = 1$.

[27] Note that in case of $\sigma < 1$, there may not even be a trade-off between private and public consumption (i.e. if υ or ξ^i are sufficiently large). In this case, individuals face a trade-off between the consumption levels of both goods on the one hand and subsequent growth on the other hand. In contrast, for $\sigma \geq 1$ there is always a trade-off between the public consumption level on the one hand and the private consumption level and growth on the other hand.

3.6 Voting equilibrium in the growth model

It is assumed that the political outcome is realized by an one-man, one-vote decision. Thus, in voting equilibrium the preferred spending fraction g^m of the individual with the median relative factor endowment ξ^m is realized. This directly follows from the facts that the individually preferred government spending fraction is monotone in the relative factor endowment and that V^i is strictly concave in g for all i. It has already been discussed in chapter 2 that it is reasonable to assume $\xi^m < 1$ and that $1 - \xi^m$ may serve as a measure of inequality.

Proposition 3.3 (Growth model): If $g^m = \tilde{g}^m$, a more equal relative factor endowment distribution, i.e. a higher median relative factor endowment ξ^m, implies a higher public consumption share g^m in voting equilibrium if $\sigma \leq 1$ or $\upsilon < 1 + \rho / \Gamma$. In case of a synthetic income tax (i.e. $\upsilon = 1$), a more equal income distribution unambiguously leads to a higher g^m.

Proof: Directly follows from lemma 3.3 and 3.4. □

3.7 Discussion

3.7.1 Inequality and growth effects of taxation

According to lemma 3.4, a more (human) capital-abundant individual may prefer a higher taxation of capital income and thus slower growth. This is in stark contrast to the usual argumentation of the inequality and growth relationship through the politico-economic channel. As already stressed, in this line of literature, capital-poor individuals demand higher capital taxes which are used to finance income transfers, thus depressing growth. (See e.g. proposition 2.2 in chapter 2). However, the analysis in this chapter has shown that this is not necessarily true if the tax revenue is used to finance public consumption. In our model, publicly provided goods and services also have a strong redistributive element since individuals consume similar amounts, although being differently taxed. Nevertheless, the voting outcome may be considerably different to the one with direct transfers. The reason for this lies in the different proposed tax schemes to be voted on. Here, due to the assumption that the factor tax rates are positively related [i.e. $\tau_K = \upsilon\tau_L$, according to (3.12)], the tax scheme implies a *positive* relationship between the net factor returns (rather than a negative one as in chapter 2). In fact, this may be more plausible. First, tax-transfer schemes in the real world rarely redistribute income towards median income earners. Rather, transfers play an important role only for poorer households. Second, income tax reforms are often such that tax rates for the median income earner and richer individuals are changed in the same direction. Tax cuts in the U.S. under the Reagan administration and tax increases after the German unification are obvious examples.

3.7.2 Welfare effects

Concerning *welfare effects* of taxation, one may consider a social planner who is concerned only with aggregate levels, as it is sometimes done in the literature on the inequality and growth relationship.[28] That is, the social welfare function is equal to indirect life-time utility of a representative individual with per capita private consumption $C(t) = C(0)e^{\vartheta t}$, where $C(0)$ is given by (3.16). How would such a social planner choose the tax rate τ for a given υ? It is easy to check that under this welfare function, the outcome would coincide with the preferred public consumption share of an individual with relative factor endowment $\xi^i = 1$. (As in chapter 2, such an individual may be called 'representative'). Hence, if $\xi^m < 1$, more equality would bring the voting equilibrium closer to the social planning optimum. In this sense, according to proposition 3.3, a more equal distribution of ξ^i may raise welfare through a *larger* public sector,[29] and thus through *higher* tax rates. This is also in contrast to standard politico-economic growth models which predict that more equality raises welfare of a representative agent through *lower* redistribution and taxes. However, as already noted in chapter 2, a representative individual only has welfare significance if lump sum redistribution of factor endowments is feasible. If it is not, the social

[28] Bertola (1993), Bénabou (1996).

[29] This result may be compared to the results about the relationship between inequality and public goods provision if public goods are privately provided. According to Itaya, De Meza and Myles (1997) and Sandler (1997), a more equal income distribution may *decrease* both the private provision level of public goods and welfare in the presence of corner solutions. Buchholz and Peters (2000) show, in addition, that an extremely skewed income distribution may even lead to a private overprovision of public goods.

planning optimum rather depends on the social welfare function. For instance, consider a Rawlsian welfare function when $l^i = L = 1$ for all i and individuals differ in capital endowments. This welfare function would be identical with the utility of the individual which is least endowed with capital rather than the one endowed with the average amount of capital. In this sense, the utility of a representative individual is biased towards owners of capital.

3.7.3 Specification of preferences and the role of savings

It may be argued that the (homothetic) preferences assumed in the growth model are somewhat special. One reason for considering these preferences is, of course, analytical tractability (i.e. to obtain steady state growth). But there are other important reasons. First, as pointed out before, the results can be compared to the literature about growth and welfare effects of redistribution policies. Second, it is argued in the following that the specification of preferences allows to compare the intertemporal (growth) model with the static model of section 3.2 in an plausible setting as far as the individual saving rates are concerned. It is interesting to work out the mechanisms underlying the different results of both models, focusing on the additional role of savings in the dynamic model.

In the static case, an increase in income of the 'decisive' voter under a majority voting rule has an ambiguous effect on his/her preferred tax-financed public consumption share. As has become clear from Fig. 3.1, on the one hand the demand for public consumption may be higher for richer individuals ('wealth effect'), whereas on the other hand the own compulsory contribution share increases ('substitution effect').

In the intertemporal context, there is an important additional mechanism. Note that in the growth model accumulated and non-accumulated factor income are distinguished. If non-accumulated factor income is taxed more heavily, the current private consumption level unambiguously decreases. In contrast, this does not have to be the case if the capital income tax rises [see (3.22) and note that $\Gamma \leq 0$ if $\sigma \leq 1$]. After an increase in the capital income tax, owners of capital adjust savings downward in order to keep up with current private consumption levels. In fact, the higher the (relative) capital endowment of an individual, the more savings decrease after a tax increase. That is, considering this *savings effect* only, an increase in the public consumption share means that the private consumption *level* of a capital-rich individual decreases less relative to the one of a capital-poor individual. In other words, capital-rich individuals are more able to adjust savings than capital-poor agents. The resulting growth reduction affects owners of capital and owners of labor similarly, since the growth rate of wages decreases in line with the capital income growth rate. Overall, only if the ratio of capital and labor taxation υ is large, such that the substitution effect of a higher relative capital endowment on the individually desired public consumption share is large, the desired public consumption share may decrease with the relative factor endowment (see lemma 3.4).

The fact that, due to the assumed preferences, the savings effect goes in the same direction as the wealth effect is an implication of the following properties of individual saving rates. First, individual saving rates rise with the relative factor endowment and, second, they decline with capital income taxation, according to (3.25). In fact, it is a well-known empirical regularity that the saving rates of households increase with individual income.[30] Hence, the specification of preferences above allows to work out a mechanism that may be quite relevant empirically.

3.8 Some empirical evidence

The theoretical analysis has revealed that a lower median to mean (capital) income differential does not necessarily yield a larger size of the government as provider of goods and services. Thus, empirically, income inequality should not be positively related to the *final government consumption share of national income*. This section provides some cross-country evidence for this hypothesis.[31]

3.8.1 Data sources

In order to control for Wagner's law, *GDP per capita* is included in some regressions.[32] Moreover, a measure for the *openness* of a country to trade, defined by the sum of exports and imports divided by GDP, is used. As Rodrik (1998) suggests, higher public expenditure reduces the volatility of income which would be especially high in relatively open economies. GDP and public consumption data are given at current prices and current

---

[30] E.g. Browning and Lusardi (1996).

[31] No effort has been made to test the hypothesis if the median voter is indeed decisive [for a discussion of this issue, see e.g. Atkinson and Stiglitz (1980, ch. 10)] or to distinguish different types of public consumption expenditures.

[32] Other tests of Wagner's hypothesis yield, at best, inconclusive results. See e.g. Bohl (1996), Gemmell (1990) and Rodrik (1998) for cross-country or time series evidence failing to support Wagner's law. Note that this is consistent with the homothetic preferences mainly analyzed in this chapter.

purchasing power parities. The sources of this data are OECD (1999a) and the Penn World Tables (PWT) 5.6.[33] Openness data are solely from PWT 5.6. Income inequality is measured by *Gini coefficients* which are taken from the World Bank (1998, Tab. 2.8) and the high quality subset of the data set by Deininger and Squire (1996).[34] Whereas a first set of regressions looks at OECD countries, a second set looks at democratic countries. This is because the median voter model should be more applicable to democracies than to, say, dictatorships.[35] A country is treated as democracy, if the Freedom House measure of democracy for a country, which derives from work of Gastil and followers (various years), lies between 1.0 and 2.5. This index is the average of a 1 to 7 scale of both a subjective index of 'political rights' and 'civil liberties', with higher ratings meaning less freedom.

3.8.2 Estimated equations and results for OECD countries

In this subsection some cross-section evidence from OECD countries is presented. For Iceland there was no inequality measure available. Some regressions also exclude the Czech Republic, Hungary, Korea, Mexico and Poland who did not join the OECD until 1994, 1995 or 1996.

In Tab. 3.1 the estimation results are shown. In the first five regressions (see columns (1)-(5) of Tab. 3.1) the average GDP per head (in thousands) and the public consumption share of the years 1994-96 from OECD (1999a) is used.[36] Moreover, the latest available Gini coefficient from the two sources given above is included. If the same latest year is available for a country, the World Bank data is used except for the third regression which uses the Deininger-Squire data to check the robustness of the results.[37] In the first

---

[33] The latter is an updated version of PWT 5 described in Summers and Heston (1991).

[34] See Deininger and Squire (1996) for the criteria for inclusion of data points in the high quality data set.

[35] Alesina and Rodik (1994) take a different view, stating that "dictators' policy decisions are also influenced by social demands and social conflicts" such that "redistributive pressures may find a political outlet not only in democracies but also in dictatorships" (p. 478). However, at least with respect to government spending for goods and services this argument does not seem to be relevant. For example, think about corruption and the extraordinary high military budgets in many non-democratic countries.

[36] Data for the Czech Republic, Hungary and Poland are in current exchange rates since purchasing power parities were not available in OECD (1999a).

[37] There are considerable differences between the World Bank and the Deininger-Squire data set with respect to Denmark, Norway and Sweden, which have much

three regressions all 28 countries are considered whereas the others only contain the 23 countries who were already OECD members in 1973.[38] In the fifth regression, openness data are averaged over 1990-92, the last years covered in the PWT 5.6.[39] In the last two regressions three year averages of GDP per capita, openness and public consumption from PWT 5.6 is used (again, to check the robustness of the results). This three year period includes the year of the latest available Gini observation before or in 1992.

	(1)	*(2)*	*(3)*	*(4)*	*(5)*	*(6)*	*(7)*
Constant	30.6***	33.44***	29.25***	35.23***	38.36***	24.7***	30.49***
	(8.782)	(7.105)	(4.944)	(5.756)	(8.573)	(3.866)	(4.271)
Gini ratio	-	-	-0.336**	-	-	-0.261*	-0.356**
	0.411***	0.443***	(-2.348)	0.463***	0.547***	(-1.821)	(-2.366)
	(-3.917)	(-3.987)		(-3.448)	(-4.640)		
GDP per	-	-0.148	-0.046	-0.155	-	-0.167	-0.206
head		(-0.900)	(0.347)	(-0.986)		(-0.905)	(-1.148)
Openness	-	-	-	-	-0.051**	-	-0.033
					(-2.800)		(-1.599)
N	28	28	28	23	23	23	23
Adj. R^2	0.347	0.342	0.118	0.313	0.482	0.057	0.125

Table 3.1: Public consumption and income inequality in the OECD.
Notes: Dependent variable is the final government consumption share of GDP. t-statistics in parenthesis. All regressions are OLS. Data sources are described in text. Levels of statistical significance are indicated by asterisks: ***99 percent; **95 percent; *90 percent.

First, contrary to Wagner's law, per capita GDP does not have a statistically significant impact on the public spending fraction of GDP (and the coefficient always has the wrong sign). Second, the data do not confirm the hypothesis by Rodrik (1998) that public consumption is used to reduce income volatility in open economies. If anything, more open OECD

higher Gini ratios in the latter source. This is likely to change estimations results since all of these countries have high public consumption shares.

[38] Before Mexico joint the OECD in 1994, the last new member has been New Zealand in 1973.

[39] For Greece and Portugal the last openness data are from 1991 and 1990, respectively.

countries seem to have *lower* government operating outlays.[40] Third, and most important, there seems to be some evidence that higher income inequality yields *lower* public consumption shares of GDP within the OECD. In the three regressions where GDP and public consumption data come from OECD (1999a) and the inequality data from the World Bank (1998) if the same latest observation is available in both sources given above, the Gini coefficient has a statistically significant negative impact on public consumption at the 99 percent level. In regression *(5)*, in which the openness variable is included, the coefficient of the Gini ratio is even significantly different from zero at the 99.9 percent level. The result of the simple regression *(1)* is illustrated in Fig. 3.2.

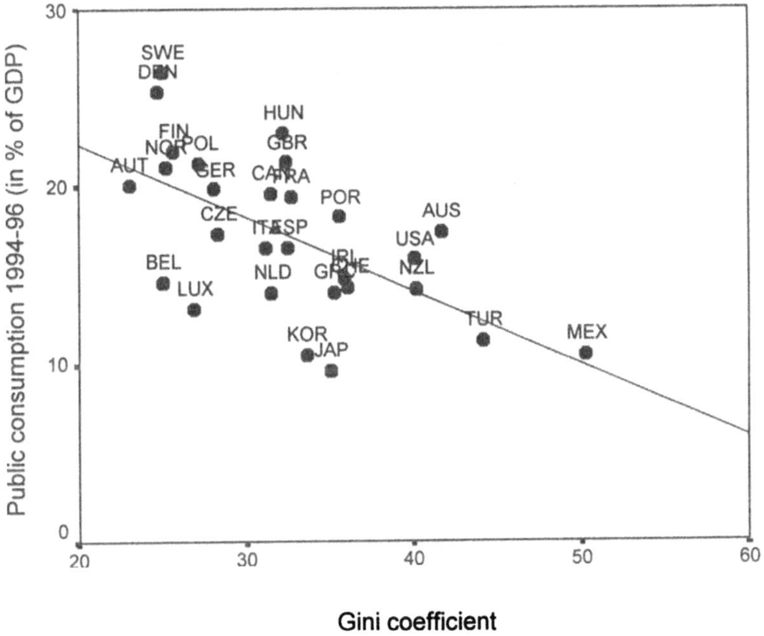

Figure 3.2: Inequality and public consumption in the OECD.

However, the results are less clear-cut if data from other sources are considered. Nevertheless, using PWT 5.6 rather than OECD (1999a) for public consumption and GDP per capita, and Deininger and Squire (1996) rather than World Bank (1998) for the Gini ratio if the same latest

[40] It should be noted that Rodrik (1998) already pointed to the fragility of his results if looking at OECD countries only.

observation is available for a country, the Gini ratio is still significant at the 90 or 95 percent level. However, the adjusted R^2 in regressions *(3)*, *(6)* and *(7)* is fairly low. But, in sum, also with these data there is some evidence for a negative relationship between inequality and the public consumption share in the OECD.

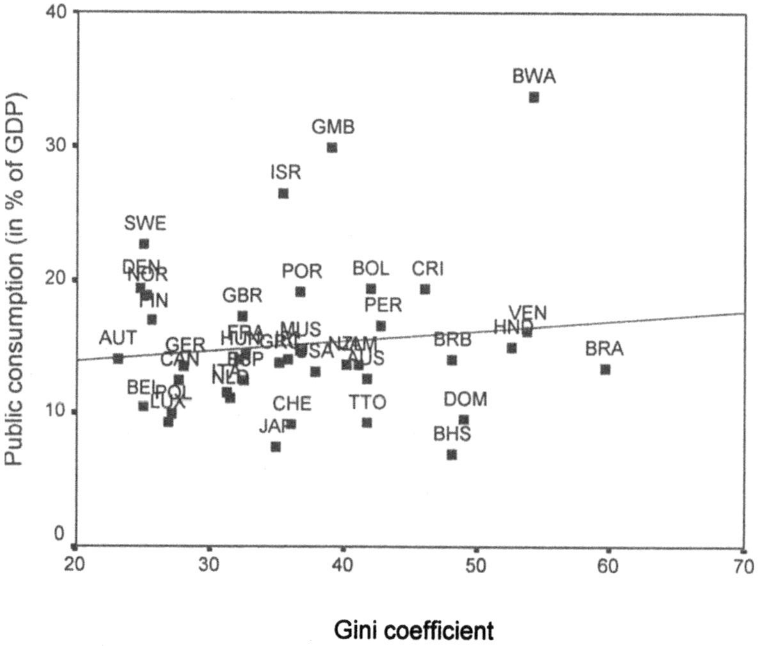

Figure 3.3: Inequality and public consumption in democracies.

3.8.3 Estimation results for democracies

However, the results are different if one looks at cross-section evidence from a larger set of countries, i.e. democracies. GDP per head, openness and final government consumption shares are three year averages from PWT 5.6, including the year of the used Gini observation. The sample consists of all democracies for which reliable Gini coefficients from the sources above for 1992 or earlier are available; that is, 24 countries presently belonging to the OECD and 15 other countries. In contrast to the estimates above, the impact of the Gini ratio on the public consumption share in all estimated equations is far from being significant.[41] (Results are not shown in detail). The same is true for openness, whereas the coefficient of GDP per head is sometimes

[41] Besides the variables used before, also regional dummies have been included.

significantly negative (!). Fig. 3.3 gives a scatter plot of the relationship between the government size and inequality similar to the one in Fig. 3.2. Although Fig. 3.3 shows a slightly positive slope, it is neither significantly different from zero nor robust.

3.9 Summary

According to a standard argument, higher income inequality fosters redistributive activities of the government in favor of the median income earner. In this chapter the relationship between income inequality and the public consumption share of total income in both a static and a dynamic median voter model has been examined. Public consumption has been assumed to be financed by (factor) income taxes. In the static case, although a higher relative income of the median voter is assumed to imply a higher individual tax share for financing public consumption, the substitution effect of this higher relative price for the publicly provided good may be dominated by a wealth effect, if the publicly provided good is normal. Thus, although public expenditure plays a redistributive role in the model, higher income inequality may nevertheless imply a smaller size of the government. Moreover, it was shown in a general equilibrium growth model, that this result may even be strengthened in a dynamic context due to different savings decisions of heterogeneous households. A dynamic model allows to distinguish between taxation effects of accumulated and non-accumulated factors of production. It has been argued that capital-rich individuals are generally better able to adjust savings than capital-poor agents. For instance, with a synthetic proportional income tax, it is optimal for capital-rich agents to reduce their current private consumption less than capital-poor individuals if the tax rate is raised. Declining savings slow down investment-driven growth. However, this growth-reducing effect of higher tax-financed public consumption affects the returns to both factors similarly. Thus, due to the saving adjustment in response to taxation, a capital-richer median voter unambiguously prefers a bigger government as provider of goods and services if the income tax is synthetic. According to a rough empirical investigation, more equal economies indeed do not seem to have smaller governments and thus are not less engaged in redistribution towards the median voter through the tax-system than unequal economies.

Appendix to chapter 3

Proof of lemma 3.1: Applying the implicit function theorem to (3.4) and manipulating the resulting expression yields

$$\text{(3.A.1)} \qquad \frac{dG^i}{dy^i}\frac{y^i}{G^i} = 1 - \frac{s'(\cdot)G^i\left(1 + MRS'(\cdot)\right)}{MRS'(\cdot)}.$$

Note that the inverse of the elasticity of substitution between the two goods can be written as $(\varepsilon^i)^{-1} = MRS'(\cdot)\frac{(y^i - s^i G^i)/G^i}{s^i}$ with $s^i = MRS(c^i/G^i)$, according to (3.4). (Also note that $c^i = y^i - s^i G^i$). Thus, we have $G^i(1 + MRS'(\cdot)) = (y^i / s^i)\varepsilon^i MRS'(\cdot)$. Substituting the latter expression into (3.A.1) yields

$$\text{(3.A.2)} \qquad \frac{dG^i}{dy^i}\frac{y^i}{G^i} = 1 - \frac{s'(\cdot)y^i}{s^i}\varepsilon^i,$$

thus confirming lemma 3.1. □

Proof of lemma 3.2: The current-value Hamiltonian function $\Im$ for the utility maximization problem of individual *i*, given his/her initial capital endowment $k_0^i > 0$ equals

$$\text{(3.A.3)} \qquad \Im(c^i, k^i, \lambda^i) = \frac{\left(c^i G^\gamma\right)^{1-\sigma} - 1}{1-\sigma} + \lambda^i(wl^i + rk^i - c^i),$$

where λ^i is the current-value shadow price of individual income of the non-accumulated factor. The first order conditions

$$\text{(3.A.4)} \qquad \frac{\partial \Im}{\partial c^i} = 0 \Leftrightarrow \lambda^i = (c^i)^{-\sigma} G^{\gamma(1-\sigma)},$$

$$\text{(3.A.5)} \qquad -\frac{\partial \Im}{\partial k^i} = \dot{\lambda}^i - \rho\lambda^i \Leftrightarrow -\hat{\lambda}^i = r - \rho$$

and the transversality condition $\lim_{t\to\infty} e^{-rt} k^i(t) = 0$ (where the latter holds if $r > \vartheta$) are necessary and sufficient for a maximum because of the concavity of $\Im$ and positive discounting (i.e. $\rho > 0$). Differentiating (3.A.4) with respect to time yields

(3.A.6) $$-\hat{\lambda}^i = \sigma \hat{c}^i - (1-\sigma)\gamma \hat{G}.$$

Combining (3.A.5) and (3.A.6) and using the expression for *r* given in (3.8) yields equation (3.20). □

Proof of lemma 3.3: Neglecting the restrictions $g \geq 0$, $g \leq \Phi$, (or) $g \leq \Phi \tau_K^{\max} / \upsilon$., the preferred spending fraction $\tilde{g}^i = \arg\max_g V^i$ is given by $\partial V^i / \partial g = 0$. It has to be shown that $\partial \tilde{g}^i / \partial \xi^i > 0$. Using (3.26), one can show with some tedious, but straightforward algebraic manipulation that $\partial V^i / \partial g = 0$ is equivalent to

(3.A.7) $$\Psi(\cdot, \xi^i, \tilde{g}^i) \equiv x(\tilde{g}^i)^2 + y\tilde{g}^i + z = 0,$$

where $z \equiv \rho\gamma\Phi^2(\rho\xi^i + \alpha a) > 0$, $x \equiv \alpha a^2(1+\gamma)\upsilon(1-\alpha) > 0$ and $y \equiv -a\Phi(1+\gamma)\big(\alpha\rho + \upsilon(1-\alpha)(\rho\xi^i + \alpha a)\big) < 0$. First note that only the smaller root of (3.A.7) can be a solution, i.e. $2x\tilde{g}^i + y < 0$. To see this, consider $\upsilon = 0$ as benchmark case. This implies $x = 0$ and thus $2x\tilde{g}^i + y = y < 0$. Hence,

(3.A.8) $$\tilde{g}^i = \left(-y - (y^2 - 4xz)^{1/2}\right) / (2x).$$

Applying the implicit function theorem and using $\partial\Omega / \partial \tilde{g}^i = 2x\tilde{g}^i + y < 0$, we have $\partial \tilde{g}^i / \partial \xi^i > 0$ if $\partial \Psi / \partial \xi^i > 0$. Using (3.A.7) and the expressions for *x*, *y* and *z*, the latter inequality holds if $\tilde{g}^i < \rho\gamma\Phi / \big(a(1+\gamma)\upsilon(1-\alpha)\big)$ which is equivalent to $-y - (y^2 - 4xz)^{1/2} < 2\alpha a \Phi \rho \gamma$, according to (3.A.8). This is further equivalent to $4xz + 4\alpha a \Phi \rho \gamma = -4\alpha\rho < 0$, confirming lemma 3.3. □

Proof of lemma 3.4: Note that $\tilde{g}^i$ increases (remains constant, decreases) with ξ^i if $\left(\partial^2 \ln c^i(0)\right)/\left(\partial g \partial \xi^i\right) > (=,<) 0$. Substituting (3.12) and (3.13) into (3.22) yields

(3.A.9)
$$c^i(0) = \frac{\Gamma(1 - \upsilon g / \Phi) + \rho}{\Omega} k_0^i + (1 - g / \Phi)\alpha a K_0 l^i,$$

and thus,

(3.A.10)
$$\frac{\partial \ln c^i(0)}{\partial g} = -\frac{\Gamma \upsilon \xi^i + \alpha a \Omega}{\left(\Gamma(\Phi - \upsilon g) + \rho\Phi\right)\xi^i + (\Phi - g)\alpha a \Omega}.$$

[Note that (3.A.10) implies $\partial^2 \ln c^i(0) / \partial g^2 < 0$, i.e. indirect life-time utility is strictly concave]. Use (3.A.10) to confirm

(3.A.11)
$$\frac{\partial^2 \ln c^i(0)}{\partial g \partial \xi^i} = \frac{\alpha a \Omega \Phi\left(\rho + \Gamma(1 - \upsilon)\right)}{\left(\left(\Gamma(\Phi - \upsilon g) + \rho\Phi\right)\xi^i + (\Phi - g)\alpha a \Omega\right)^2}.$$

(i) First, consider $\sigma > 1$ which implies $\Gamma > 0$. [Remember $\Gamma = (\sigma - 1)(\gamma + 1)(1 - \alpha)a$]. Thus, the right hand side of (3.A.11) is positive (zero, negative) if $\upsilon < (=, >) 1 + \rho / \Gamma$.
(ii) Second, note that $r = (1 - \alpha)a(1 - \tau_K) > \vartheta$ implies $\rho > -\Gamma(1 - \tau_K)$, according to (3.21), and that $\tau_L \leq 1$ is equivalent to $\upsilon = \tau_K / \tau_L \geq \tau_K$. Since $\sigma < 1$ implies $\Gamma < 0$, $\rho > -\Gamma(1 - \tau_K)$ implies $\rho > -\Gamma(1 - \upsilon)$. This concludes the proof. □

Chapter 4. Skilled labor reallocation, wage inequality, and growth

4.1 Motivation

As we have seen in chapter 1 (section 1.5), despite a rising relative supply of high-skilled labor, wage inequality across skill groups has surged in the U.S. and Great Britain, and, at least, it has not declined in most other countries.[1] In addition, both absolute and relative unemployment rates of low-skilled labor have substantially increased in almost all OECD countries.[2] These facts strongly indicate that relative labor demand has shifted towards high-skilled labor, thus deteriorating (at least relative) earning opportunities of low-skilled workers. According to a standard view, this shift has been due to "a change in the production function that raises the marginal product of the skilled relative to the unskilled" [Krugman (1994, p. 37)], a hypothesis which is commonly referred to as 'skill-biased technological change' (SBTC).[3]

In this chapter a different technology-related channel to explain the observed labor market developments is suggested. It is argued that a reallocation of high-skilled labor from production to *skill-intensive non-production activities* changes relative labor demand if high-skilled and low-skilled labor are technological complements in production; that is, a decrease in high-

---

[1] E.g. Gottschalk and Smeeding (1997).

[2] E.g. Nickell and Bell (1997, Tab. 10.2), OECD (1997b, Tab. 4.1b).

[3] For a discussion of the theoretical and empirical literature on SBTC, see chapter 1 (section 1.5). Another explanation is that increased import competition of products which are intensive in low-skilled labor mainly contributed to relative labor demand shifts [e.g. Freeman (1995)]. However, according to tests based on the Heckscher-Ohlin-Samuelson model [e.g. Hanson and Harrison (1995), Neven and Wyplosz (1996)] and factor content studies [e.g. Katz and Murphy (1992)] the evidence for the trade hypothesis is rather weak. Moreover, employment shifts towards skilled labor in developed countries has taken place mainly within rather than between firms and industries (see section 1.5 for a discussion of empirical evidence and references), contrary to what is expected if increased trade in final goods mainly contributed to a shift in relative labor demand. However, looking at trade in intermediate goods (i.e. inputs for the production of consumption goods) rather than final goods, a fall in the relative price of inputs in developed countries reduces unskilled labor demand if intermediate goods and unskilled labor are good substitutes. In this case, skill upgrading would take place even within industries which do not produce traded goods [see e.g. Aghion, Caroli and García-Peñalosa (1999)].

skilled labor in production decreases the marginal productivity of low-skilled labor in production. Non-production activities are assumed to increase the total factor productivity and are specified as R&D. The model reflects the notion that technological shocks (like the wide-spread adoption of personal computers and new information technologies) lead to productivity gains *only if* (high-skilled) labor resources are shifted towards the search of new ways to organize and coordinate work. Thus, contrary to the standard view, *technology-induced changes in relative labor demand do not have to be reflected in changes of the relative productivity of different types of labor*. This difference to the hypothesis of SBTC is of considerable importance for both theoretical and empirical reasons. First, if technological change would be skilled labor-saving, i.e. would make high-skilled workers more efficient in *production* activities already performed by high-skilled workers (sometimes called 'intensive' SBTC),[4] then the effect on relative labor demand would be ambiguous. More precisely, if the production technology is represented by a neoclassical production function with high-skilled and low-skilled labor, skilled labor-augmenting technical progress raises relative demand for high-skilled workers *only if* the elasticity of substitution between both types of labor exceeds unity. In appendix 4.A, this is shown for a linear homogenous neoclassical production function with high-skilled and low-skilled labor as only inputs.[5] In contrast, if technological change leads to a reallocation of high-skilled workers towards skilled labor-intensive non-production activities, relative demand for high-skilled labor in production rises *unambiguously* (at given wages). Second, as has been extensively discussed in chapter 1 (see section 1.5), a reallocation of (mainly) high-skilled workers towards non-production activities is exactly in line with the empirical evidence on the skill-bias hypothesis. However, this kind of evidence has been commonly interpreted as relative labor productivity shift for *given* production employment levels. This theoretical view has been criticized since it is hardly testable.[6] In contrast, our model is not only empirically supported but also provides a more differentiated view *why* high-skilled labor has shifted to non-production activities [see also Grossmann (2000b)].

In order to fix ideas, a segmented labor market for high-skilled and low-skilled workers is introduced in the endogenous growth framework of Aghion and Howitt (1992). In this framework, production efficiency

---

[4] See Johnson (1997) for a discussion of 'intensive' versus 'extensive' SBTC. According to his definition, extensive SBTC is making the skilled more efficient in tasks formerly performed by the unskilled, i.e. is not skilled-labor saving.

[5] See Vogt (1999) for further discussion.

[6] E.g. Blanchard (1997), Thurow (1998).

improves in a Hicks-neutral way, i.e. it leaves the relative productivity parameters of high-skilled and low-skilled labor unchanged. There are two production-related activities and one non-production-related activity in the economy. The non-production activity is specified as search for new intermediate products.[7] For simplicity, only the latest blue-print for an intermediate good (which is associated with the highest total factor productivity) is produced and used as input for the production of a homogenous consumption good. Regarding the technology, two crucial assumptions drive the results of the model. First, the non-production (i.e. research) activity is assumed to be skill-intensive (for simplicity, *only* high-skilled labor is used). Second, in the final goods production high-skilled and low-skilled labor are technological complements.

The special framework of Aghion and Howitt (1992) provides an appropriate structure for the general idea of the paper. However, as will be argued later, it is not necessary for the results and is merely chosen for its familiarity.

The basic mechanism of the model is as follows. If the expected gains from the non-production activity (i.e. research incentives) increase, high-skilled labor is reallocated from production towards non-production (i.e. R&D).[8] Because of the decreasing employment of high-skilled labor in production, the relative marginal productivity of low-skilled labor declines. Hence, if wages are fully flexible, the wage differential between high-skilled and low-skilled labor rises due to the resulting shift in relative labor demand.

However, if relative wages are sticky, low-skilled employment may be declining as well.[9] In the model, *relative wage stickiness* arises from

---

[7] This structure is similar to the endogenous growth model of Romer (1990) which could also be applied to formalize the hypotheses of this chapter. The basic differences between the model of Romer (1990) and the model of Aghion and Howitt (1992) are the following. First, in the latter model innovations arrive according to a stochastic process whereas they are deterministic in the former. Second, in contrast to the Romer-model, in the Aghion-Howitt-model old intermediate goods are replaced when new innovations arrive.

[8] In the adopted framework of Aghion and Howitt (1992), intermediate goods production is monopolized due to patent rights. Thus, increasing research incentives stimulate skilled labor demand in R&D by raising profits of the incumbent monopolist.

[9] Labor market outcomes with rigid wages are often discussed under the assumption of a homogenous labor force. However, as Blinder (1988, p. 4) points out: "Once we force ourselves to think seriously about the *heterogeneity* of labor, the very concept of wage rigidity loses precision. For example, is it the *average level* of wages or the structure of *relative* wages that is sticky?" (italics original).

efficiency wage payments due to *social comparisons between high-skilled and low-skilled workers* in final goods production.[10] Following Akerlof and Yellen (1988, 1990), workers adjust their effort level downward whenever they feel treated unfairly. Thus, firms can improve work effort by paying higher wages.[11] As a new feature of the model also *within-group social comparisons* of (low-skilled) workers are considered. As a result, if the wage rate of the low-skilled in final goods production exceeds its marginal productivity, the wage rate of the low-skilled in intermediate goods production may not adjust sufficiently downward to imply full employment. That is, within-group social comparisons give rise to a kind of *absolute wage stickiness*. In general, labor rationing (i.e. due to efficiency wage payments) in one part of the economy is not sufficient for unemployment to occur. Rather, a necessary condition for unemployment is wage inflexibility throughout the entire economy.[12] To my knowledge, the interaction between efficiency wage considerations in different sectors have not been explored yet.[13]

Although unemployment could also arise due to minimum wages or other wage rigidities, the reason for using a sociological efficiency wage model is threefold. First, relative unemployment rates of low-skilled workers are more or less similar across the OECD, suggesting that labor market regulations are negligible for understanding *low-skilled* unemployment.[14] For instance, even in flexible labor markets like the U.S. and Great Britain, unemployment of low-skilled (male) workers has risen substantially since the early 1970s, and neither its change nor its level is, on average, lower

---

As it turns out, since there is more than one type of labor in our model, relative wage rigidity alone is not a sufficient condition for unemployment.

[10] These social comparisons may be due to envy, status or equity concerns. See also the discussion in chapter 2 of this book.

[11] The literature on labor market effects of this kind of 'gift exchange' between employers and employees has been pioneered by Akerlof (1982).

[12] In contrast, Falkinger (2000) shows that wage inflexibility is generally *not* necessary for involuntary unemployment to occur.

[13] Most work on efficiency wage theory focuses on one-sector models. An exception is Agénor and Aizenman (1997) who, however, only assume efficiency wage payments in one sector. In the other sector of their model, wages do not flexibly adjust due to institutional rigidities.

[14] As Falkinger and Grossmann (1999) argue, low-skilled employment crucially depends on the quality of primary education whereas differences in *overall* OECD unemployment rates are affected by wage rigidities. However, as pointed out by Schlicht (1992), fair wages may also depend on bargained wages suggesting a channel through which institutional rigidities and fairness considerations interact.

than in other OECD countries.[15] *Non*employment rates for low-skilled males (which also include unemployed workers out of the labor force) in the U.S. and Great Britain currently lie well above 30% (!).[16] Moreover, evidence on the widely agreed hypothesis that minimum wages are an important cause of low-skilled unemployment is, to say the least, inconclusive.[17] Second, recent studies provide overwhelming empirical evidence for sociological efficiency wage models and their implications for wage stickiness. These include surveys on the motivating forces of decision-makers within firms behind the payment structure,[18] results about fairness considerations in experimental studies[19], and evidence for the hypothesis that job satisfaction of workers crucially depend on reference/comparison wage levels.[20] Moreover, there is evidence that wages of low-skilled workers positively vary with wages of better qualified workers within industries.[21] Third, as motivated earlier, our set-up allows to analyze the interaction between social comparisons within and across wage groups.

The analysis suggests three kinds of relationships: (i) between technological change (i.e. growth) and employment of low-skilled workers, (ii) between technological change and wage equality, and (iii) between low-skilled employment and wage equality.[22] Unlike in Saint-Paul (1996), Agénor and Aizenman (1997) and Gregg and Manning (1997), the derived relationships between technological change, wage inequality and unemployment do not rest on changes in relative labor productivity parameters.

The chapter is organized as follows. Section 4.2 sets up an efficiency wage model with production and non-production activities performed by low-skilled and high-skilled workers, respectively. Section 4.3 defines the steady

---

[15] E.g. Nickell and Bell (1996).

[16] Nickell and Bell (1997), Murphy and Topel (1997). According to Cohen, Lefranc and Saint-Paul (1997), *non*employment rates of low-skilled males in France are considerably lower than in the U.S. whereas *un*employment rates of this group are similar in both countries.

[17] E.g. Machin and Manning (1997).

[18] E.g. Levine (1993), Agell and Lundborg (1995), Bewley (2000).

[19] E.g. Fehr, Kirchsteiger and Riedl (1993), Fehr, Gächter and Kirchsteiger (1996).

[20] Clark and Oswald (1996). See also chapter 2 for a discussion of happiness studies.

[21] Slichter (1950), Krueger and Summers (1987, 1988).

[22] For the literature on the growth-employment relationship, see e.g. Pissarides (1990, ch. 2), Bean and Pissarides (1993), Hoon and Phelps (1997), Aghion and Howitt (1994), Falkinger and Zweimüller (2000). In contrast, our main focus is on wage inequality, relative employment rates and technological change rather than on growth and aggregate employment. See section 4.5 for further discussion of the relation of our model to this literature.

state equilibrium and derives the central mechanism underlying the basic hypothesis of this chapter with respect to the observed relative labor demand shift towards qualified workers. Comparative static results are presented in section 4.4. Section 4.5 discusses the basic ideas of the model and relates it to empirical evidence. The last section summarizes. All proofs as well as the transitional dynamics towards the steady state equilibrium are given in appendix 4.B-4.E. Moreover, appendix 4.A looks at the underlying forces of a change in the relative marginal productivity of labor in a segmented labor market, if there is skilled labor-saving technological progress and labor supply is inelastic. The notion of skilled labor-saving technological progress plays an important role in the literature on SBTC. Appendix 4.A reveals that a bias in labor demand in favor of skilled workers only occurs if the elasticity of substitution between skilled and unskilled labor exceeds unity.

4.2 The model

There is a segmented labor market for high-skilled ("skilled") and low-skilled ("unskilled") labor. Unskilled labor can be allocated to produce an intermediate good which is an input for the production of a homogenous consumption good. Skilled labor used in non-production raises the expected value of total factor productivity in the final goods production. This non-production activity can be interpreted as research for new intermediate products. Besides the latest (i.e. most productive) intermediate good, both skilled and unskilled labor are inputs for the production of the consumption good. The supply of both skilled and unskilled labor is inelastic and denoted by H and L, respectively. However, *effort* supply of workers (rather than the supply of working hours) are functions of actual wages *relative to the workers' reference wages*. These reference wages are determined by social comparisons. Labor markets are perfectly competitive in the sense that firms take marginal costs per unit of *effective* labor services as given. The economy is closed and agents are assumed to have perfect foresight. Both final goods production and research takes place in perfect competition, whereas the intermediate goods production can be monopolized by a successful innovator.

Denote t as the number of innovations which have already occurred. Time is continuous and the length between two innovations t and t+1 is random. Let τ_t denote the point in time in which innovation t arrives and σ_t the random waiting time until the next innovation t+1 arrives. It is assumed that the waiting time σ_t for innovation t+1 to occur is exponentially distributed with parameter λ_t. This implies that the number of events which occur in a certain time interval is 'Poisson-distributed',[23] i.e. the probability that n innovations occur within a time interval of length $d\tau$ is given by $(\lambda_t d\tau)^n \exp(-\lambda_t d\tau)/(n!)$.[24] Thus, the probability that innovation t+1 does *not* arrive in time interval $d\tau$ is given by $\exp(-\lambda_t d\tau)$. (To see this, set $n = 0$).[25] Each monopoly lasts only until the next innovation is made, whereas each patent for an intermediate product lasts forever. The price for each patent is equal to the expected present value of the flow of monopoly profits. Denote the instantaneous profit of the incumbent monopolist of patent t as π_t and the price of patent t as V_t. If π_t is constant during the time interval $[\tau_t, \tau_t + \sigma_t]$, then the expected discounted value of profits of an monopolist with an exogenous discount rate r is given by $V_t = \int_{\tau_t}^{\infty} \pi_t \exp\left(-(r+\lambda_t)(\tau - \tau_t)\right) d\tau = \pi_t / (r + \lambda_t)$. (Note that it has been accounted for the fact that $\exp(-\lambda_t d\tau)$ is the probability that an incumbent monopolist keeps his/her monopoly during any time interval $d\tau$).

4.2.1 Final goods production

Output Y of the consumption good (which is the numeraire commodity) is produced with skilled and unskilled labor as well as with an intermediate product of quantity x. The technology of a representative firm exhibits constant returns to scale and is, at each instant, given by the following production function:

$$Y = A \cdot (e^{HY} H^Y)^{\alpha} (e^{LY} L^Y)^{\beta} x^{1-\alpha-\beta}, \tag{4.1}$$

where $\alpha > 0$, $\beta > 0$, $\alpha + \beta < 1$. H^Y and L^Y denote the amounts of skilled and unskilled labor with corresponding effort levels e^{HY} and e^{LY}, respectively. Following Aghion and Howitt (1992), the productivity parameter A increases with a constant factor $\gamma > 1$ each time an innovation occurs (γ may be called the 'size of innovation'), i.e.

---

[23] E.g. Billingsley (1979).

[24] The expected number of innovations in time interval $d\tau$ is equal to $\lambda d\tau$.

[25] It should be noted that, according to the model considered in this chapter, the parameter λ of the Poisson distribution of the number of innovations changes during the transition to the steady state each time an innovation occurs. However, what we need is the probability that *no* innovation arrives in time interval $d\tau$ after t innovations occurred. This probability is indeed equal to $\exp(-\lambda_t d\tau)$.

(4.2) $$A_t = A_0 \gamma^t.$$

(Remember that t is the number of innovations which have already occurred rather than time). $A_0 > 0$ is assumed to be historically given.

4.2.2 Intermediate goods production

As will be shown below, all variables remain at the same level during the random time interval $[\tau_t, \tau_t + \sigma_t]$ in which the monopolized production of patent t lasts. Let H_t^Y, L_t^Y, e_t^{HY} and e_t^{LY} denote the levels of skilled and unskilled employment and effort provision, respectively, in final goods production within that time interval. According to (4.1), the tth monopolist of an intermediate product faces an inverse demand function

(4.3) $$p_t(x) = (1-\alpha-\beta) A_t (e_t^{HY} H_t^Y)^{\alpha} (e_t^{LY} L_t^Y)^{\beta} x^{-(\alpha+\beta)}.$$

Intermediate goods are produced according to a constant returns to scale technology

(4.4) $$x = e^{LX} L^X,$$

where $(e^{LX} L^X)$ are efficiency units of unskilled labor employed by the incumbent monopolist. Let w^{LX} denote the wage rate of unskilled labor in intermediate goods production, i.e., according to (4.4), w^{LX} / e^{LX} are marginal costs per unit of output (i.e. the marginal costs per unit of effective labor services). Thus, maximization of instantaneous profits of the tth monopolist implies

(4.5) $$x_t \equiv \arg\max_{x \geq 0} \left\{ p_t(x) x - (w_t^{LX} / e_t^{LX}) x \right\} = \left(\frac{A_t (1-\alpha-\beta)^2 (e_t^{HY} H_t^Y)^{\alpha} (e_t^{LY} L_t^Y)^{\beta}}{(w_t^{LX} / e_t^{LX})} \right)^{\frac{1}{\alpha+\beta}}.$$

Hence, the incumbent monopolist sets a price $p = (w^{LX} / e^{LX}) / (1-\alpha-\beta)$ as fixed mark-up over marginal costs per unit of effective labor.[26] (For

[26] To see this, substitute (4.5) into (4.8).

notational convenience, the variable index t is suppressed whenever this does not lead to confusion). Moreover, as long as the intermediate product has not become obsolete yet, instantaneous profits are given by

$$\text{(4.6)} \qquad \pi = (\alpha+\beta)px = A\left(\frac{a(e^{HY}H^Y)^\alpha(e^{LY}L^Y)^\beta}{\left((w^{LX}/e^{LX})/A\right)^{1-\alpha-\beta}}\right)^{\frac{1}{\alpha+\beta}},$$

where $a \equiv (\alpha+\beta)^{\alpha+\beta}(1-\alpha-\beta)^{2-\alpha-\beta} > 0$ is an unessential constant.

4.2.3 Research

Remember that the waiting time σ_t for innovation t+1 to occur is exponentially distributed with parameter λ_t (implying that innovations are governed by a Poisson process). As argued in e.g. Aghion and Howitt (1998, p. 55), λ_t approximately equals the instantaneous probability that innovation t+1 occurs. Assume that this probability is given by $\lambda_t = f(e_t^{HR}H_t^R)$, where H_t^R is the amount of skilled labor used in research after t innovations, e_t^{HR} is the corresponding effort level and $f(\cdot)$ is an increasing and concave function with $f(0)=0$. Thus, the more effective skilled labor is devoted to research, the higher is the probability that a monopoly is replaced in any given time interval.

Let w_t^{HR} denote the wage rate of skilled labor in research after t innovations. The representative research unit maximizes the flow of expected profits of innovation t+1 in any point of time. V_{t+1} denotes the expected discounted value of monopoly profits (and thus the 'price') of innovation t+1. Thus, the problem $\max_{H^R}\left\{f(e_t^{HR}H^R)V_{t+1} - w_t^{HR}H^R\right\}$ s.t. $0 \le H^R \le H$ is solved. Hence, in an interior solution the labor input is chosen such that the expected marginal product of one efficiency unit of skilled labor equals its wage per unit of effort:[27]

$$\text{(4.7)} \qquad w_t^{HR}/e_t^{HR} = f'(e_t^{HR}H_t^R)V_{t+1}.$$

[27] A sufficient condition for an interior solution of the allocation of skilled labor in perfect foresight equilibrium derived below is given in appendix 4.D.

Since all variables (i.e. employment levels, wages, and effort levels) are unchanged while a certain patent is in operation, instantaneous monopoly profits π_{t+1} are constant within the random time interval $[\tau_{t+1}, \tau_{t+1} + \sigma_{t+1}]$. Thus, as derived above, the value of the $(t+1)$st innovation is given by

$$V_{t+1} = \frac{\pi_{t+1}}{r + f(e_{t+1}^{HR} H_{t+1}^{R})}. \tag{4.8}$$

(Remember that the discount rate r is exogenous).[28] The denominator in (4.8) can be interpreted as the interest rate faced by the $(t+1)$st innovator incorporating the risk of being replaced by the next innovation.

4.2.4 Effort supply and fair wages

Effort supply in either group is assumed to be an increasing and concave function of the actual wage w paid by firms relative to a reference wage $\tilde{w}$; that is

$$e^i = \tilde{e}(w^i / \tilde{w}^i), \; i = HY, LY, LX, HR. \tag{4.9}$$

Reference wages have the interpretation as wage rates workers perceive as fair. (These fair wages will be specified below). If actual wages are equal to fair wages, workers supply a normalized effort equal to unity; i.e., $\tilde{e}(1) = 1$. Firms choose wages and labor inputs in order to maximize profits subject to the respective effort supply functions (4.9), taking fair wages $\tilde{w}$ as given. It is easy to show that wages are set according to the following rule:

$$\frac{\tilde{e}'(w^i / \tilde{w}^i)(w^i / \tilde{w}^i)}{\tilde{e}(w^i / \tilde{w}^i)} = 1, \; i = HY, LY, LX, HR. \tag{4.10}$$

---

[28] Treating the interest rate as exogenous in an intertemporal model may seem peculiar. This can be theoretically founded by assuming that the intertemporal utility of each individual equals the discounted life-time consumption (with discount rate r) and there either is a frictionless Walrasian credit market or no credit market at all. As it is argued in Aghion and Howitt (1992, p. 331), if there is no credit market "all nonresearch workers consume their wage income at each instant, the owners of the monopoly intermediate firm consume their flow of profits at each instant, and research workers receive no pay unless their firms innovate, at which time they are paid in shares of the next intermediate firm".

(4.10) states that it is optimal for firms to pay wages such that the elasticity of effort supply with respect to the ratio of actual to fair wages equals unity. This is similar to the 'Solow-condition' (Solow, 1979; Schlicht, 1978), and is explicitly stated in Schlicht (1992). (4.10) implies that wages are set proportionally to their reference levels.
Reference wages are assumed to be based on social comparisons among workers and are specified as follows. In final goods production, fair wages are given by[29]

(4.11) $$\tilde{w}^{LY} = \eta w^{HY} + (1-\eta) w^{L*}, \ 0 \le \eta \le 1,$$

(4.12) $$\tilde{w}^{HY} = \nu w^{LY} + (1-\nu) w^{H*}, \ 0 \le \nu \le 1,$$

where w^{L*} and w^{H*} denote the market clearing wages of unskilled and skilled labor, respectively. That is, the fair wage of unskilled (skilled) workers in final goods production is the weighted average of the wage received by skilled (unskilled) coworkers and the overall market clearing wage of unskilled (skilled) workers. If there would not be any social comparisons among workers, w^{L*} and w^{H*} would be the economy's full employment equilibrium wages.[30] However, as will be seen below, due to fairness considerations of unskilled workers, there may neither be full employment among the unskilled nor may wages of unskilled labor in final and intermediate goods production be equal. Note that according to (4.11) and (4.12), if $\eta < 1$ and $\nu < 1$, wage levels perceived as fair depend on labor market conditions,[31] i.e. fair wages are low when productivity and thus market clearing wages are low.[32] However, fair wages may depart from market clearing levels because of social comparisons *across skill groups* (i.e. $\eta > 0$ and $\nu > 0$).

---

[29] See Akerlof and Yellen (1990) for a similar specification in a model with one representative firm.

[30] That is, without social comparisons, w^{L*} and w^{H*} would be equal to the marginal productivity of unskilled and skilled workers in final goods production, respectively.

[31] For empirical evidence that labor market conditions substantially affect workers' fairness perceptions, see e.g. Kahneman, Knetsch and Thaler (1986).

[32] As shown in appendix 4.B, w^{L*} positively depends on the equilibrium employment level of skilled labor in production H^Y and negatively on labor supply L.

Concerning fair wages of both unskilled workers in intermediate goods production and skilled workers in research, it is assumed that these workers make *within-group* comparisons with workers in the final goods production. This may be formalized as

$$\text{(4.13)} \qquad \tilde{w}^{LX} = \mu w^{LY}, \ \mu > 0,$$

$$\text{(4.14)} \qquad \tilde{w}^{HR} = w^{HY}.$$

According to (4.13) and (4.14), both reference wages are proportional to the actual wages of equally qualified workers in final goods production. As we focus on unskilled employment, fair wages of skilled workers in research are not only proportional but also equal to skilled wages in production.[33]
In order to distinguish skilled and unskilled workers in a meaningful way, it is plausible to assume the following.

Assumption 1: Skilled workers are either sufficiently more productive or sufficiently short in supply relative to unskilled workers such that $w^{H*} > w^{L*}$ holds for any allocation of skilled workers in production and non-production.

Now let us specify the effort supply function (4.9). If the actual wage paid to workers is lower than the fair wage, effort supplied by workers is only a fraction of its normal level. Accordingly,[34]

$$\text{(4.15)} \qquad e^i = \tilde{e}(w^i / \tilde{w}^i) = \min\{(w^i / \tilde{w}^i), 1\}, \ i = HY, LY, LX, HR.$$

Under this specification, the wage setting rule (4.10) holds for any wage $w \leq \tilde{w}$, since the marginal costs per unit of effective labor in either group equals its wage per unit of effort w/e. Moreover, if $w \leq \tilde{w}$, w/e equals the fair wage $\tilde{w}$, according to (4.15). In order to obtain unique wages, it is sufficient to impose the following.

---

[33] This is assumed here merely for simplicity reasons. Moreover, one would not gain any further insight if allowing reference wages in (4.13) and (4.14) to be some weighted average analogous to (4.11) and (4.12).
[34] This follows Akerlof and Yellen (1990) who argue (giving a extensive list of references), that this specification formalizes a "fair wage-effort hypothesis" which is consistent with various sociological theories.

Assumption 2: If profits are unaffected by paying different wages including the wage workers perceive as fair, firms prefer to pay the fair wage.[35]

Lemma 4.1: Firms do not pay less than fair wages, i.e. workers supply normal effort $e^i = 1$, $i = HY, LY, LX, HR$.

All lemmas are proven in appendix 4.B.

Lemma 4.2: We have $w^{LY} = \eta w^{H*} + (1-\eta)w^{L*}$ and $w^{HR} = w^{HY} = w^{H*}$, i.e. skilled labor is fully employed.

Regarding skilled labor, $w^{H*} > w^{L*}$ (according to assumption 1) and lemma 4.2 imply that $w^{H*} > w^{LY}$ if $\eta < 1$. Thus, if $\eta < 1$ and $\nu > 0$, (4.12) implies that $\tilde{w}^{HY} < w^{H*}$, i.e. fair wages of skilled workers are below market clearing. Thus, equilibrium wages may exceed fair wages, if paying exactly fair wages would lead to excess labor demand. If $\eta = 1$, then $\tilde{w}^{HY} = w^{H*} (= w^{LY})$. In any case, the fairness parameter ν does not affect equilibrium wages. Regarding unskilled labor, $w^{H*} > w^{L*}$ and lemma 4.2 imply that wages of unskilled workers in final goods production lie above the overall market clearing level if and only if these workers make social comparisons, i.e.

(4.16) $$w^{LY} = w^{L*} \text{ if and only if } \eta = 0,$$

(4.17) $$w^{LY} > w^{L*} \text{ if and only if } \eta > 0.$$

To deal with wage inequality, it is useful to look at relative wages. Note that according to lemma 4.2, the relative wage of skilled labor in final goods production is given by

(4.18) $$\frac{w^{H*}}{w^{LY}} = \left(\eta + (1-\eta)\frac{w^{L*}}{w^{H*}}\right)^{-1} \geq 1,$$

---

[35] Alternatively to (4.15) and assumption 2, one may assume a strictly concave effort supply function to obtain unique optimal wages. For example, using $\tilde{e}(w / \tilde{w}) = 1 + \log(w / \tilde{w})$ rather than (4.15) would leave all results unaffected.

where the latter inequality is due to assumption 1 [(4.18) holds with equality if and only if $\eta = 1$]. Moreover, since firms take wages per unit of effort as given and $e^{HY} = e^{LY} = 1$ (according to lemma 4.1), (4.1) implies that

$$(4.19) \qquad \frac{w^{H*}}{w^{LY}} = \frac{\alpha L^Y}{\beta H^Y}.$$

From (4.18), it becomes clear that the higher η, the less relative wages are determined by market forces, and the more by fairness considerations. Thus, if $\eta > 0$ relative wages in final goods production are *sticky*, i.e. compressed in the sense that relative wages of skilled workers are below relative marginal productivity. However, relative wage stickiness is not sufficient for unemployment of unskilled workers. This is because unskilled workers who are not employed in final goods production could also be employed in intermediate goods production. Remember that $w^{LX} \geq \tilde{w}^{LX} = \mu w^{LY}$, according to lemma 4.1 and (4.13). Thus, if within-group social comparisons are sufficiently strong, i.e. μ is sufficiently high, there is excess supply and thus unemployment of unskilled labor. In order to focus the analysis, the following assumption ensures that this is always the case if relative wages are sticky.

Assumption 3: If relative wages are sticky (i.e. if $\eta > 0$), then $\mu > \mu^*$, where μ^* is defined as the largest value of μ at which all unskilled workers not employed in final goods production would find employment in intermediate goods production.

Lemma 4.3: If $\eta > 0$ or $\mu \geq 1$, we have $w^{LX} = \mu w^{LY}$. If $\eta > 0$ or $\mu > 1$, there is unemployment among the unskilled. If $\eta = 0$ and $\mu \leq 1$, we have $w^{LX} = w^{L*} = w^{LY}$ and thus full employment of unskilled labor.[36]

[36] Note that μ^* is endogenous and we have $\mu^* < 1$ if $\eta > 0$, according to the definition of μ^*. As will be directly implied by the results in section 4.4, in steady state equilibrium μ^* rises with *r* and *H*, falls with γ, *L* and η, and does not depend on ν. Note that if $\eta > 0$ and $\mu \geq 1$, assumption 3 is not necessary for unemployment.

Note that if $\eta = 0$ and $\mu < 1$, equilibrium wages of unskilled workers in intermediate goods production exceed their fair wages since otherwise there would be excess demand for unskilled labor. In the following, the analysis is focused on the case $\eta > 0$, i.e. relative wages are sticky and there is unemployment of unskilled labor.

Remark: Note that $w^{HR} = w^{HY} = w^{H*}$ and, if $\eta = 0$ and $\mu \leq 1$, $w^{LY} = w^{LX} = w^{L*}$, according to lemma 4.2 and lemma 4.3, respectively, Thus, if $\eta = 0$ and $\mu \leq 1$, all results of our model would be the same as if there would not be any social comparisons (or efficiency wage considerations, respectively) in the model. Of course, since there would not be unemployment of unskilled workers in this case, only the relationship between wage inequality and technological change could be analyzed. But *qualitatively*, all results with respect to relative wages would be the same as in the case with social comparisons, as long as $\eta < 1$ (i.e. relative wages are not fully rigid).

4.3 Perfect foresight growth equilibrium

In perfect foresight equilibrium wage rates of skilled labor in research and production are equal and always adjust until the market for skilled labor is cleared (i.e. we have $w_t^{HY} = w_t^{HR} = w_t^{H*}$, according to lemma 4.2). This defines a sequence $(H_0^Y, H_1^Y, H_2^Y, \ldots)$, where $H_t^Y = H - H_t^R$ for all $t \geq 0$. (Recall that the variable index t denotes the period of random length σ_t in which the tth innovation is monopolized). The allocation of skilled labor in production and research determines all wage levels, the employment levels of unskilled labor in intermediate and final goods production as well as the economy's rate of growth simultaneously. In this section, it is derived how unskilled employment and relative wages are affected by the allocation of skilled labor in production and research. Moreover, the balanced growth equilibrium is defined. As derived in section 4.4, the impacts of the parameters H, γ, and r on the steady state allocation of skilled labor then imply how H, γ, and r affect relative wages, employment levels and growth in steady state. As it is also shown in section 4.4, the parameters L, η and μ affect employment and relative wages through a different channel than through the impact on the skilled labor allocation.

4.3.1 How the allocation of skilled labor affects relative wages and unskilled employment

Denote the unemployment rate of unskilled labor as

$$u^L = 1 - \frac{L^X}{L} - \frac{L^Y}{L}. \tag{4.20}$$

Lemma 4.4:

(i) If $\eta < 1$, relative wages of skilled workers decrease with the amount of skilled labor used in production, i.e. $\frac{\partial (w^H / w^{LY})}{\partial H^Y} < 0$ and $\frac{\partial (w^H / w^{LX})}{\partial H^Y} < 0$.

(ii) If $\eta > 0$, the employment level of unskilled labor in final goods production increases with the amount of skilled labor used in production, i.e. $\frac{\partial L^Y}{\partial H^Y} > 0$.

(iii) If $\eta > 0$, the unemployment rate of unskilled labor decreases with the amount of skilled labor used in production, i.e. $\frac{\partial u^L}{\partial H^Y} < 0$.[37]

The intuition of lemma 4.4 is the following. A higher amount of skilled labor devoted to production raises the relative marginal product of unskilled labor since skilled and unskilled labor are technological complements in final goods production, i.e. $(\partial^2 Y)/(\partial H^Y \partial L^Y) > 0$, according to (4.1). Hence, if $\eta < 1$, relative wages decline unambiguously with H^Y [part (i) of lemma 4.4]. (If $\eta = 1$, relative wages are solely determined by fairness considerations and thus are *totally* rigid). If $\eta > 0$, relative wages are sticky due to social comparisons of unskilled workers in final goods production. This implies that an increase in skilled labor allocated to production unambiguously reduces the unemployment rate of the unskilled [part (iii) of lemma 4.4]. The reason for this is as follows. First, employment of unskilled workers in final goods production L^Y rises due to their increased marginal

---

[37] Note that lemma 4.4 (ii) implies that μ^* positively depends on H^Y if $\eta > 0$.

productivity [part (ii) of lemma 4.4]. Second, as shown in the proof of lemma 4.4 (see appendix 4.B), labor demand of the incumbent monopolist and thus employment in intermediate goods production L^X is unambiguously positively related to the employment level of unskilled labor in the final goods production. Intuitively, if a higher amount of unskilled labor is employed in final goods production, the inverse demand function (4.3) faced by the incumbent monopolist shifts upwards. Hence, the marginal revenue of the incumbent monopolist increases. Moreover, (4.1) implies that wages and employment in final goods production are negatively related for both skilled and unskilled labor. Thus, according to lemma 4.3, an increase in L^Y lowers marginal costs (i.e. wages w^{LX}) in intermediate goods production, unambiguously raising labor demand.[38]

Remark: If $\eta = 0$, then relative wages are fully flexible. Thus, in this case there are no employment effects of a change in the allocation of skilled labor, even if there is unemployment among the unskilled (i.e. even if $\mu > 1$, in addition to $\eta = 0$, according to lemma 4.3).

4.3.2 Balanced growth

Define a balanced growth equilibrium as a stationary solution for the productivity-adjusted wage rate $\omega_t^{H*} = (w_t^{H*} / A_t)$ of skilled workers, productivity-adjusted wages $\omega_t^{LX} = (w_t^{LX} / A_t)$, $\omega_t^{LY} = (w_t^{LY} / A_t)$ of unskilled workers, and employment levels H_t^R, $H_t^Y (= H - H_t^R)$, L_t^X and L_t^Y. These steady state equilibrium values are denoted $\hat{\omega}^{H*}$, $\hat{\omega}^{LX}$, $\hat{\omega}^{LY}$, $\hat{H}^R$, $\hat{H}^Y$, $\hat{L}^X$ and $\hat{L}^Y$ (with a corresponding unemployment rate $\hat{u}^L$), respectively.
A formal analysis of the transitional dynamics is given in appendix 4.C. As in Aghion and Howitt (1992), in transition to the steady state there is a negative relationship between current and future use of skilled labor in production. According to (4.6), (4.8) and $H_{t+1}^Y = H - H_{t+1}^R$, a foreseen increase in the amount of skilled labor in production next period raises the expected value of the next innovation, because of both rising monopoly

[38] In appendix 4.B it is also shown that that the elasticity of the unskilled employment level in the final goods production with respect to skilled labor lies between zero and unity. Moreover, it is easy to show that this elasticity increases with η, the degree of which unskilled workers socially compare themselves with skilled workers in final goods production.

profits next period and a lower risk of being displaced by a new monopolist. Hence, according to (4.7) and lemma 4.2, *current* wages of skilled workers increase, in turn depressing current demand for skilled workers in final goods production. Also note that according to lemma 4.4, the model predicts both unskilled employment levels and relative wages to fluctuate during the transition to the steady state.

The average growth rate of final output $Y_t = \gamma Y_{t-1}$ is endogenously determined by the amount of skilled labor used in research. As can be shown analogously to Aghion and Howitt (1992, p. 336), the average steady state growth rate of the economy is given by

$$g = f(\hat{H}^R)\ln\gamma . \tag{4.21}$$

Thus, growth increases with both the steady state amount of skilled labor devoted to research $\hat{H}^R$ and the factor γ by which total factor productivity increases each time an innovation occurs. (Note that $\gamma > 1$ has been assumed).[39] Note that due to the assumption $f(0) = 0$, if the entire skilled labor force would be used in production, there would be no output growth in this economy.

4.4 Comparative static results

Concerning the steady state allocation of skilled labor and growth, neither the introduction of a segmented labor market for skilled and unskilled labor nor allowing for efficiency wages in the basic framework of Aghion and Howitt (1992) qualitatively affect their results. (See appendix 4.D for a formal derivation). First, a higher discount rate r reduces the expected discounted value of an innovation, making research less profitable relative to production. This causes the amount of skilled labor used in research $\hat{H}^R$ to fall, which in turn has a negative impact on the average growth rate, according to (4.21). Second, the size of innovations γ positively affects future profit streams in intermediate goods production and thus $\hat{H}^R$, unambiguously yielding faster growth. Third, a rising skilled labor supply H has a negative impact on wages of skilled labor in both research and final

[39] As also shown in Aghion and Howitt (1992), output evolves as a random walk with (positive) drift in this type of model. The variance of the growth rate in (4.21) equals $f(\hat{H}^R)(\ln\gamma)^2$.

goods production, yielding higher skilled labor demand for both non-production and production activities.
In the following, comparative static results of an interior steady state equilibrium, in which strictly positive amounts of skilled labor are devoted to both research and final goods production, are derived. All results are summarized in Tab. 4.1.[40]

CHANGE IN: EFFECT ON:	r	γ	H	L	$H, L^{\oplus}$	η	μ
$\hat{H}^Y$	+	–	+	0	+	0	0
$\hat{H}^R$	–	+	+	0	+	0	0
$\hat{u}^L$	–	+	–	+	–	+	+
$\hat{\omega}^H / \hat{\omega}^{LY}$	–	+	–	+	–	–	–
$\hat{\omega}^H / \hat{\omega}^{LX}$	–	+	–	+	–	–	–
$\hat{\omega}^{LY} / \hat{\omega}^{LX}$	0	0	0	0	0	0	–
$\hat{\omega}^H$	–	+	–	+	–	–	–
$\hat{\omega}^{LY}$	+	–	+	–	+	+	–
$\hat{\omega}^{LX}$	+	–	+	–	+	0	+
g	–	+	+	0	+	0	0

Table 4.1: Comparative static results of an interior steady state equilibrium if $0 < \eta < 1$. $^{\oplus}$ $\Delta H + \Delta L = 0$.

4.4.1 Change in research incentives

First, consider parameter changes which directly affect expected monopoly profits and thus future values of innovations, e.g. by lower discount rates or larger productivity gains per innovation.

Proposition 4.1: Consider an interior steady state equilibrium with $0 < \eta < 1$. Both a decrease in the discount rate r and an increase in the size

[40] For the interested reader, Tab. 4.1 also contains the results for the absolute wage levels which are not derived in the following in order to focus the discussion.

of innovations γ raises *(i)* the unskilled unemployment rate $\hat{u}^L$, *(ii)* relative wages $\hat{\omega}^H/\hat{\omega}^{LY}$ and $\hat{\omega}^H/\hat{\omega}^{LX}$, and *(iii)* the average growth rate g.

All propositions are proven in appendix 4.E.[41]

A lower discount rate r raises the average growth rate of the economy by raising the steady state amount of skilled labor in research (see above). The same is true for an increase in the size of innovations γ, which, in addition, also has a direct impact on growth, according to (4.21). Concerning unemployment and relative wages, the mechanisms underlying lemma 4.4 apply: As a lower amount of skilled labor is devoted to production, relative marginal productivity of unskilled labor in final goods production declines. However, due to fairness considerations between both skill groups, relative wages of unskilled labor do not fall to the same degree. Hence, even though the wage gap between skilled and unskilled workers widens, the unskilled employment level in final goods production is reduced. For given within-group social comparisons, the resulting downward shift of the inverse demand function faced by the incumbent monopolist also lowers employment in intermediate goods production.[42]
From this it becomes clear that also *policies* designed to raise private incentives to research may have these effects. Thus, policies like R&D subsidies would have to be implemented carefully. If, for example, in this model R&D subsidies would be financed by progressive income taxation (in the sense that the tax rate rises with the level of wages), relative wages would become less sticky *if* social comparisons between wage groups would be based on *after-tax* wages. This may counter the negative employment effects of skilled labor reallocation towards research. However, financing R&D subsidies with a flat income tax would unambiguously raise unemployment by increasing the amount of skilled labor in research without affecting reference wages.

4.4.2 Change in labor supply

Now consider a change in the supply of both skilled and unskilled labor, e.g. through educational training programs or immigration.

---

[41] For a discussion of related literature concerning the relationship between growth and unemployment, see the next section.

[42] The net effect on real wages paid to unskilled workers is ambiguous since aggregate productivity rises, but productivity-adjusted wages decline (see Tab. 4.1).

Proposition 4.2: Consider an interior steady state equilibrium with $0 < \eta < 1$.

(i) An increase in skilled labor supply H reduces the unskilled unemployment rate $\hat{u}^L$, lowers relative wages $\hat{\omega}^H / \hat{\omega}^{LY}$ and $\hat{\omega}^H / \hat{\omega}^{LX}$, and raises the average growth rate g.

(ii) An increase in unskilled labor supply L raises the unskilled unemployment rate $\hat{u}^L$, raises relative wages $\hat{\omega}^H / \hat{\omega}^{LY}$ and $\hat{\omega}^H / \hat{\omega}^{LX}$, and does not affect the growth rate g.

First, a positive share of an additional skilled labor supply will be employed in production. Hence, again lemma 4.4 applies; that is, both unskilled unemployment and wage inequality is reduced because of an increase in relative marginal productivity of unskilled labor.[43] Moreover, as a larger amount of skilled labor is devoted to research, the economy is growing faster in its steady state. Second, wage inequality increases with L since unskilled workers in final goods production consider it fair to receive lower wages when labor supply of their skill group increases.[44] This leads to a higher employment level of unskilled labor in final goods production. Because also fair wages in intermediate goods production decline due to within-group comparisons, labor demand increases here as well. However, because of wage compression due to fairness considerations (see subsection 4.2.4), relative wages are not flexible enough to keep the unskilled unemployment rate constant.[45]

Concerning the allocation of skilled labor, an increase in unskilled labor supply is neutral in steady state equilibrium with the technology in (4.1). This is because marginal productivity of skilled labor in research and production, respectively, is equally increased by a rising unskilled employment level. Generally, the net effect of rising unskilled labor supply on the allocation of skilled labor and thus on growth is ambiguous. The analysis suggests that educating or training unskilled workers is even more effective in reducing both unskilled unemployment and wage gaps between

[43] For a discussion of related literature concerning the impact of an increase in the supply of skilled labor on relative labor demand, again, see the next section.

[44] To see this, note that the hypothetical market clearing wage w^{L*} decreases with L, according to equation (4.B.4) in appendix, implying a fall in fair wages since $\tilde{w}^{LY} = w^{LY} = \mu^{-1} \tilde{w}^{LX} = \mu^{-1} w^{LX}$ positively depends on w^{L*} if $\eta < 1$, according to (4.11)

[45] In other words, the elasticity of unskilled employment with respect to unskilled labor supply lies between zero and unity, as can be formally derived from equations (4.B.3) and (4.B.6) in appendix.

skilled and unskilled workers than immigration of skilled labor, since the former also reduces the unskilled labor force. However, this comparative static result may be misleading as education is costly in the real world.[46]

4.4.3 Change in fairness considerations

Fairness considerations of unskilled workers have three effects in the model. First, firms are induced to pay wages which lead to unemployment of unskilled labor in order to elicit high effort. Second, social comparisons in final goods production create relative wage stickiness, preventing wages to adjust sufficiently to keep unskilled unemployment at a constant rate, when (relative) earning prospects of unskilled workers deteriorate. Third, within-group social comparisons among the unskilled lead to absolute wage stickiness.

Proposition 4.3: Consider an interior steady state equilibrium with $0 < \eta < 1$. The unskilled unemployment rate $\hat{u}^L$ rises and relative wages $\hat{\omega}^H / \hat{\omega}^{LY}$ and $\hat{\omega}^H / \hat{\omega}^{LX}$ fall with fairness parameters η and μ, respectively.

More intensive social comparisons of unskilled workers with skilled workers in final goods production (i.e. a higher η) ceteris paribus raise wages of the unskilled and thus reduces employment of unskilled labor in final goods production. In turn, this induces relative wages of skilled labor to fall, according to (4.19). Moreover, labor demand of the incumbent monopolist also declines. The reason for this is the following. First, due to within-group comparisons wage demands in the intermediate goods production are positively related to wages of the unskilled in final goods production where the latter rise unambiguously. This raises marginal costs of the incumbent monopolist. Second, a decreasing employment level in the final goods production shifts the inverse demand function of the incumbent monopolist downward, in turn reducing marginal revenues. Hence, unskilled unemployment unambiguously increases with η. A higher μ implies lower labor demand in intermediate goods production by raising marginal costs of the incumbent monopolist. Hence, the production of intermediate goods declines. This in turn reduces unskilled labor demand in final goods production since a decreasing input of the intermediate good lowers the

---

[46] Some countries, e.g. Australia and Canada, select immigrants on basis of their education to some extent. In Germany, the discussion about such a policy has just started.

marginal productivity of unskilled labor. Thus, because of its decreasing employment, relative marginal productivity of unskilled labor increases, inducing relative wages to fall.

With the Cobb-Douglas technology (4.1), fairness parameters η and μ do not have any impact on the steady state allocation of skilled labor. According to the discussion above, the marginal productivity of skilled labor in both final goods production and research declines with η and μ, respectively, yielding lower skilled labor demand in both units. With respect to the allocation of skilled labor, both effects exactly cancel in balanced growth equilibrium. It should be noted, that fairness perceptions in intermediate and final goods production are mutually reinforcing, because of the vertical integration of both production activities. That is, fairness considerations in one production activity affects both labor demand and wages in the other one. In contrast, social comparisons of skilled labor in final goods production (represented by the parameter ν) do not have any impact on wages or employment as skilled workers always receive market clearing wages.

4.5 Discussion

4.5.1 How the comparative-static results relate to the literature

As already argued in chapter 1 and the motivation of this chapter, a negative relationship between technological change and the relative demand for low-skilled labor has been derived by hypothesizing a change in relative productivity parameters. This kind of productivity shift (i.e. SBTC) has either been formulated as exogenous [e.g. Saint-Paul (1996), Agénor and Aizenman (1997), Gregg and Manning (1997)] or made endogenous [e.g. Galor and Tsiddon (1997), Acemoglu (1998)]. In our model, the derived relationship between (growth-promoting) technological change and relative labor demand (i.e. an increase in both wage inequality and unskilled unemployment) is due to a change in the allocation of skilled labor in production and non-production activities (where the latter has a positive impact on growth) rather than a skill-biased shift in the production function. But besides suggesting a new view about relative labor demand shifts, the model offers new channels through which both innovations (and thus growth; see proposition 4.1) and education (i.e. the supply of skills; see proposition 4.2) affects unemployment. This also should be related to the literature.

Both a positive and negative relationship between growth and employment has been theoretically suggested. Taking the rate of productivity growth as exogenous, Pissarides (1990, ch. 2) and Hoon and Phelps (1997) derive a positive impact of growth on employment when there are matching frictions. In both models, higher expected productivity growth raises expected discounted profits of firms from creating new vacancies which raises employment.[47] Aghion and Howitt (1994) show that this effect may, however, be dominated by a creative destruction (i.e. 'business stealing') effect of innovations, which works in the opposite direction, since plants using old technologies may be driven out of the market. Rather than investigating the impact of growth on unemployment, Bean and Pissarides (1993) suggest that higher unemployment reduces savings and thus investment-driven growth. In Falkinger and Zweimüller (2000) growth is innovation-driven and education is necessary for employment. Employment and the rate of growth are simultaneously determined by factors affecting the education of workers and the incentives to innovate. For instance, educational subsidies boost both employment and growth, whereas transfers financed by labor income taxes have the opposite effect.

In Falkinger and Zweimüller (2000), the relationship between education and growth is positive. In contrast, Saint-Paul (1996) considers a model with matching frictions, in which an increase in relative supply of skilled workers reduces the employment rate of unskilled workers. This is because the likelihood for firms to fill a vacant job with a skilled rather than a less productive unskilled worker rises with the supply of skills. Gregg and Manning (1997) suggest that an increase in the relative supply of skilled labor by government intervention would be effective to counter the adverse effects of SBTC on the demand for low-skilled labor.

In our model, there is a trade-off between endogenous technical change and employment due to social comparisons among workers. Moreover, skill-upgrading reduces unskilled unemployment due to a decline in relative wages of skilled workers, in turn relaxing wage compression. The latter effect is similar to the one in Gregg and Manning (1997) who, however, attribute unemployment to bargaining power of workers.

4.5.2 Skilled labor reallocation versus biased change in production

The model predicts that a reallocation of skilled labor away from activities which are directly related to production results in a reduction of the demand

[47] In addition, Hoon and Phelps (1997) show that in an open economy (i.e. for a fixed interest rate), a decreasing ratio of non-wage to wage income dampens the workers' quit rates, thus increasing employment further.

for unskilled labor, thus fostering wage inequality and unemployment. In fact, there is plenty of evidence from OECD countries for a substantial increase in the non-production employment share within most industries in the last two or three decades. This has been extensively discussed already in chapter 1 (section 1.5). For example, to remind the reader, Berman, Bound and Machin (1998) find an annualized change of 0.3 percentage points for the U.S. non-production employment share during the 1980s, where 73% of this change has been within industries. This labor reallocation towards non-production activities is found to be highly correlated with an increase of skilled labor within industries.[48] Moreover, it has been shown that this kind of skill-upgrading is significantly correlated with both the R&D expenditure[49] and computer-intensity of an industry.[50] Also the share of the total labor force devoted to R&D has increased sharply in the developed world. For instance, according to Jones (1995, Fig. 1), in 1988 almost one million people have been employed in R&D in the U.S which are nearly twice as many as in the early 1970s.

One may object that the special consideration of research as non-production activity in our model does neither reflect the *total* increase of the non-production employment share nor accounts for the fact that the shift mainly occurred *within* industries. However, the structure of the model allows a broader interpretation.

The crucial part of the model is that qualified workers in non-production search for and realize productivity-enhancing ways to *organize* production and work (i.e. in order to raise A). This fits well into the notion that major innovations like the so-called 'computer revolution' have induced firms to be engaged in substantial restructuring.[51] The basic idea is that for the adoption of recent information technology to increase total factor productivity, resources had to be shifted towards activities that increase the flow of information within a firm and to find new ways to coordinate, supervise and train workers in order to reap the gains from job rotation and team work, and so on.[52] In terms of our model, such an exogenous

---

48 E.g. Berman, Bound and Grichilis (1994), Machin, Ryan and van Reenen (1996).

49 E.g. Machin and van Reenen (1998).

50 E.g. Autor, Katz and Krueger (1998).

51 See e.g. Lindbeck and Snower (1996), who however do not distinguish between production and non-production activities. For case studies indicating substantial organizational changes, see e.g. Hammer and Champy (1993).

52 For example, computerization allows to create large operational databases which make it profitable to be engaged in research on the effectiveness of workers, the payment morale of customers, inventory control, and so on. Moreover, large databases of transactions help marketing managers to react to changes in customer

technology shock can be viewed as an increase in γ, fostering organizational changes with respect to the skilled labor allocation. Thus, although the model analyzes the interaction between *research* and total factor productivity, the mechanism underlying the reallocation effect is very likely to play an important role for other non-production activities as well. Are the results and the structure of the model at odds with the empirical finding that the shift towards non-production employment has mainly occurred within firms and industries? In contrast to the theoretical literature on SBTC (but in line with the empirical one), the new explanation for relative labor demand shifts in the model has been derived by analytically distinguishing between production and non-production activities. Generally, non-production activities are supposed to enhance productivity rather than to produce physical output. One may imagine that there are non-production units within firms or that firms have 'outsourced' those business services, e.g. to consulting or outside research firms. Although the research unit in the endogenous growth model of Aghion and Howitt (1992) used in this paper is usually interpreted as outside research sector, it should be noted that Aghion and Howitt (1998, p. 437) interpret the research activity in their model also as knowledge production *within* firms, stating that

> the amount of resources devoted to the creation of knowledge is certainly underestimated by standard measures of R&D. [...] Many workers [...] in management or other nonresearch activities spend considerable amount of their time and energy in looking for better ways of producing and selling the output of the enterprise they are employed by, and hence their compensation should be counted [...] as part of the cost of creating knowledge.

This is exactly in line with our basic hypothesis. For the results of our model it only matters that non-production units are separated from production units. If this is the case, the marginal productivity of low-skilled workers in production units is affected when skilled labor is shifted towards non-production. Thus, relative labor demand shifts in the model are consistent with both non-production employment shifts within and between firms and industries, respectively. Does the distinction of non-production units within and outside a firm matter for the role of social comparisons in the model? Since we do not consider social comparisons of similarly educated workers

---

demands. [Bresnahan (1999)]. As Snower (1999a) points out, new information technologies permit firms to design and manufacture products through a parallel process in which design engineers and manufacturing employees interact. Compared to an organization in which design and manufacturing is a sequential process, such an organizational structure is likely to increase both the gains from designing products and the productivity of *all* employees.

within the same unit, it is unessential if social comparisons among workers are made within or between firms, as long as they are made *across units*. In other words, it is the analytical distinction of units rather than firms that matters for the results of the model.

4.5.3 Productivity growth and skilled labor reallocation

According to the discussion above, the model suggests that the wide-spread adoption of information and communication technologies has opened up new possibilities to raise productivity which, however, could only be realized by shifting high-skilled workers to non-production activities. We have shown that this, in turn, adversely affects low-skilled labor demand. According to (4.2) and (4.21), if skilled labor is reallocated due to an increase in γ (rather than a decrease in r), both output growth and factor productivity growth is raised.[53] One may argue that this is (yet) not supported by the data. However, since according to our basic hypothesis an increase in γ is technology-related, one may fall back on the extensive discussion about the 'productivity puzzle' reported in the literature on information technologies (i.e. that productivity growth does not seem to have accelerated despite the 'computer revolution'). First, it has been forcefully argued that "it takes time and resources for the potential productivity of the new technology to be fully realized" [Ahn (1999, p. 5)]. It may even be the case that "new technologies are, in fact, initially broadly inferior to older technologies they seek to replace", although they most likely will "ultimately dominate older systems of production across a wide variety of activities" [Young (1993, p. 446)]. Although these 'learning costs' are, for simplicity, not reflected in the model, the notion that resources have to be provided in order to reap the benefits of these technologies is exactly in line with our basic hypothesis. In fact, recent evidence for the U.S. suggests that productivity growth eventually has surged considerably in the second half of the 1990s.[54] Second, many authors

---

[53] In the model, output growth is also promoted by an increase in non-production employment, according to (4.21). As extensively discussed in the endogenous growth literature, this is due to a 'scale effect' (i.e. an increase in R&D input raises long-run growth). However, the presence of such a 'scale effect', embodied in the model of Aghion and Howitt (1992) which is used here for its familiarity and simplicity, is unessential for all of our results. As Young (1998), Segerstrom (1998) and Aghion and Howitt (1998, ch. 12) have shown, the 'scale effect' can be eliminated without affecting any other basic predictions of innovation-driven growth models. See Jones (1999) for further discussion.

[54] European Commission (1999), WIFO (1999), Economist (2000, p. 51).

believe that, due to the surge in the use of information equipment, the long discussed problems of measuring the growth of output and productivity have become much more severe [e.g. Grilichis (1994), Aghion and Howitt (1998), OECD (1999b)].[55] For instance, Aghion and Howitt (1998, ch. 12) show that even in a steady state with balanced growth the failure of national account statistics to reflect the substantial improvements in the quality of both capital and consumption goods contributes to a severe underestimation of the rates of productivity and output growth.[56]

4.5.4 Social comparisons and unemployment

According to Agell and Lundborg (1995, p. 302) survey evidence from manufacturing firms suggests that "blue-collar workers apparently compare wages both within their own firm and across firms". The structure of our model allows to analyze the interaction between these two types of social comparisons. If $\eta > 0$, wages are compressed in final goods production due to social comparisons of unskilled workers with skilled workers. However, the analysis has shown that this is not sufficient to create unemployment. Only if within-group social comparisons of unskilled workers in intermediate goods production are sufficiently strong (i.e. μ is sufficiently large) such that wages of the unskilled in this production unit does not fully adjust, there is unemployment among the unskilled.[57] In other words, wage

[55] Zvi Grilichis even devoted his presidential address at the AEA congress in 1994 to this issue, noting that "our ability to interpret changes in aggregate total factor productivity has declined, and major portions of actual technical change have eluded our measurement framework entirely. [...] and thus its productivity effects, which are likely to be quite real, are largely invisible in the data. That there were gains is not really in doubt. Just observing the changes in the way banks and airlines operate, and in the ways in which information is delivered to firms and consumers, would lead one to conclude that we are in the midst of a major technological revolution." [Grilichis (1994, p. 10-11)].

[56] With respect to other measurement problems like the failure to account for output of knowledge (which, e.g., results from non-production activities), as investment (as it is done for physical capital), Aghion and Howitt (1998, p. 435-436) conclude: "These measurement problems particularly distort standard measures of growth during a period of *transition* such as we are going through now when the information revolution has greatly enhanced the opportunities for knowledge creation. In particular, they imply that GNP and productivity may appear to be *slowing down* when in fact they are *surging*." (italics added).

[57] In Aghion and Howitt (1992) only skilled labor is productive in intermediate goods production. In this case, $\eta > 0$ would be a sufficient condition for unemployment. Thus, with this specification the labor market effects of within-

compression in final goods production creates wage pressure in intermediate goods production and within-group social comparisons imply that absolute wage levels are rigid. Moreover, a reallocation of skilled labor away from final goods production reduces labor demand in both final *and* intermediate goods production for given fairness perceptions of workers. From this one can conclude that in the case of related (e.g. vertically integrated) production activities, skilled labor reallocation towards non-production activities even affects labor demand in parts of the economy in which this kind of restructuring does not take place.

4.6 Summary

It has been shown that incentives to reallocate high-skilled labor from production to non-production (i.e. research) activities depresses relative demand for low-skilled workers if skilled and unskilled workers are technological complements in production. The basic idea is that in order to reap the benefits of major technological changes, like the adoption of personal computers and information technologies, industries have to restructure which means that resources (i.e. skilled workers) have to be shifted. This accounts for the fact that the non-production employment share and an observed reorganization of firms has been found to be highly correlated with a deterioration of the relative position of low-skilled workers. Since the relative marginal productivity of labor changes only because of relative employment effects, this is a different argument than the usual hypothesis of SBTC, commonly viewed as an ‘‘exogenous shift in the production function’’ [Berman, Bound and Machin (1998, p. 1250)]. Whereas the usual skill-bias hypothesis is not directly testable, our view is supported by the fact that there has been a substantial increase of skilled non-production workers, especially in R&D-intensive and computer-intensive industries. Moreover, whereas relative labor demand effects of skill-augmenting shifts in a neoclassical production function crucially depend on the elasticity of substitution between skilled and unskilled labor (as shown in appendix 4.A), the reallocation effect suggested in this chapter

---

group social comparisons could not be addressed. Another reason behind our modification of incorporating unskilled labor in (4.4) is the notion that once a patent is invented, production itself only requires routine tasks. Assuming alternatively that both types of labor can be employed in intermediate goods production would not affect the main results of this paper, since it is the allocation of skilled labor between production and non-production activities which is essential in the model.

unambiguously reduces unskilled labor demand. With firms paying efficiency wages based on social comparisons among workers, the resulting labor market effects depend on the interaction between relative wage stickiness, arising from social comparisons of the unskilled with skilled workers, and absolute wage stickiness, arising from within-group social comparisons among unskilled workers. It has been shown that wage compression due to social comparisons of workers across skill groups may only result in unemployment of unskilled workers, if unskilled workers make within-group social comparisons as well.

Appendix to chapter 4

Appendix 4.A. The relative marginal productivity of labor

Assume there is a representative firm producing a homogenous good of quantity Y with skilled and unskilled labor. Let the production function F be linear homogenous, concave and increasing in its arguments, i.e.

$$(4.A.1) \qquad Y = F(Bh, l) \equiv l f(B\kappa),$$

where h, l is the quantity of skilled and unskilled labor, respectively, $f(\cdot) \equiv F(\cdot, 1)$ and $\kappa \equiv h / l$. $B > 0$ is a productivity parameter, and an increase in B represents skilled labor-saving technological progress. Note that $f'(\cdot) > 0$ and $f''(\cdot) \leq 0$. The ratio of the marginal productivity of skilled and unskilled labor, denoted ω, is given by

$$(4.A.2) \qquad \omega \equiv \frac{\partial F / \partial h}{\partial F / \partial l} = \frac{B f'(B\kappa)}{f(B\kappa) - B\kappa f'(B\kappa)}.$$

It is easy to show that ω is a non-increasing function of κ. The elasticity of substitution between both types of labor is equal to $\sigma \equiv -(\partial \kappa / \partial \omega)(\omega / \kappa)$. Thus, the inverse of the elasticity of substitution is given by

$$(4.A.3) \qquad \sigma^{-1} = -\frac{\partial \omega}{\partial \kappa} \frac{\kappa}{\omega} = -\frac{B\kappa f(B\kappa) f''(B\kappa)}{f'(B\kappa)\left(f(B\kappa) - B\kappa f'(B\kappa)\right)},$$

according to (4.A.2). Now suppose that all variables are time dependent. Let a dot over a variable denote its derivative with respect to time and a head denote its rate of change. Using (4.A.2), the rate of change of the relative marginal productivity is given by $\hat{\omega} = \frac{\left(B(\dot{B}\kappa + B\dot{\kappa}) f'' + \dot{B} f'\right)\left(f - B\kappa f'\right) + B(\dot{B}\kappa + B\dot{\kappa}) f' f''}{B f'\left(f - B\kappa f'\right)}$. Rearranging terms yields

$$(4.A.4) \qquad \hat{\omega} = \hat{B} + \frac{B\kappa f f''}{f'\left(f - B\kappa f'\right)}(\hat{B} + \hat{\kappa}) = \hat{B}\left(1 - \frac{1}{\sigma}\right) - \frac{\hat{\kappa}}{\sigma},$$

where (4.A.3) is used for the latter equation. Thus, if B increases, relative marginal productivity, and thus, if the representative firm takes wages as given, relative labor demand changes in favor of skilled workers if and only if $\sigma > 1$. Let H, L denote the inelastic supply of skilled and unskilled labor, respectively. If wages are flexible, we have $\kappa = H / L$ in equilibrium and ω is the wage differential between both skilled and unskilled labor. Thus, all other things equal, market forces depress the wage differential if the relative skill supply rises, i.e. if $\hat{\kappa} > 0$.

Appendix 4.B. Proof of lemmas 4.1-4.4

Proof of lemma 4.1: According to (4.10) and (4.15), all firms are indifferent in paying any wage rate $w^i \in (0, \widetilde{w}^i]$, $i = HY, LY, LX, HR$. Thus, the result directly follows from assumption 2. □

Proof of lemma 4.2: The lemma is proven in three steps.
Step 1: According to lemma 4.1 and (4.14), we have $w^{HR} \geq \widetilde{w}^{HR} = w^{HY}$. Suppose $w^{HR} > w^{HY}$. Then the representative unit could unambiguously gain by cutting wages down to $w^{HR} = w^{HY}$ since $e^{HR} = 1$ whenever $w^{HR} \geq w^{HY}$. Thus $w^{HR} = w^{HY}$.
Step 2: Suppose $w^{HR} = w^{HY} < w^{H*}$. According to the definition of w^{H*} as overall market clearing wage, this would create excess demand for skilled labor. Hence, competition in the labor market implies $w^{HR} = w^{HY} \geq w^{H*}$.
Step 3: Note that $w^{LY} \geq \widetilde{w}^{LY}$, according to lemma 4.1, and $w^{HR} = w^{HY} \geq w^{H*}$, according to step 2. Using (4.11), this implies $w^{LY} \geq \widetilde{w}^{LY} = \eta w^{HY} + (1-\eta) w^{L*} \geq \eta w^{H*} + (1-\eta) w^{L*}$. Now suppose $w^{HR} = w^{HY} > w^{H*}$, which would result in unemployment for skilled labor. Thus, underbidding of unemployed skilled workers in final goods production would also lower fair wages of unskilled workers in final goods production. Thus, competition would drive wages down to $w^{HY} = w^{H*}$ and skilled employment levels up to full employment. Moreover, since there is no excess demand for unskilled labor at fair wages, and full effort is provided whenever $w^{LY} \geq \widetilde{w}^{LY}$, we have $w^{LY} = \widetilde{w}^{LY} = \eta w^{H*} + (1-\eta) w^{L*}$. □

Proof of lemma 4.3: Remember that $w^{LX} \geq \widetilde{w}^{LX} = \mu w^{LY}$, according to lemma 4.1 and (4.13). If $\eta > 0$, then there is unemployment of unskilled labor, according to assumption 3. Moreover, if $\eta = 0$, then $w^{LY} = w^{L*}$, according to (4.16). Thus, if $\eta = 0$ and $\mu > 1$, then $w^{LX} \geq \tilde{w}^{LX} = \mu w^{L*} > w^{L*}$, and thus there is unemployment among the unskilled. If $\eta = 0$ and $\mu = 1$, there would be full employment at fair wages $\tilde{w}^{LX} = w^{LY} = w^{L*}$, and $w^{LX} \geq \tilde{w}^{LX} = w^{L*}$ must hold in labor market equilibrium. Since firms do not pay more than fair wages if there is no excess labor demand at fair wages, we have $w^{LX} = w^{LY} = w^{L*}$ and full employment if $\eta = 0$ and $\mu = 1$. Similarly, if $\eta > 0$ or $\mu \geq 1$, there is no excess labor demand which implies $\tilde{w}^{LX} = w^{LX} = \mu w^{LY}$. Only the case $\eta = 0$ and $\mu < 1$ remains to argued. If $\eta = 0$ and $\mu < 1$, then there would be excess demand of unskilled labor at fair wages $\tilde{w}^{LX} = \mu w^{LY} = \mu w^{L*} < w^{L*}$ such that $w^{LX} > \tilde{w}^{LX}$ must hold in labor market equilibrium. Since there is no excess demand if $w^{LX} \geq w^{L*}$ and workers provide full effort if $w^{LX} = w^{L*}$, we have $w^{LX} = w^{LY} = w^{L*}$. □

Proof of lemma 4.4: Using (4.1) and lemma 4.1, the wage rate of unskilled workers in the final goods production becomes

$$\text{(4.B.1)} \qquad w^{LY} = A\beta(H^Y)^{\alpha}(L^Y)^{\beta-1}x^{1-\alpha-\beta} = A\left(\frac{b\left(H^Y/L^Y\right)^{\alpha}}{(w^{LX}/A)^{1-\alpha-\beta}}\right)^{\frac{1}{\alpha+\beta}},$$

where $b \equiv \beta^{\alpha+\beta}(1-\alpha-\beta)^{2(1-\alpha-\beta)} > 0$, and the latter equation is due to substitution of (4.5) and the use of lemma 4.1. Substituting $w^{LX} = \mu w^{LY}$ from lemma 4.3 into (4.B.1) yields

$$\text{(4.B.2)} \qquad w^{LY} = Ab\mu^{-(1-\alpha-\beta)}\left(H^Y/L^Y\right)^{\alpha} \text{ and } w^{LX} = Ab\mu^{\alpha+\beta}\left(H^Y/L^Y\right)^{\alpha}.$$

With (4.4) and lemma 4.1 it follows that $L^X = x$. Hence, substituting (4.B.2) into (4.5), one obtains a positive relationship between unskilled labor in the intermediate and the final output sector as suggested in subsection 4.3.1:

$$\text{(4.B.3)} \qquad L^X = (c/\mu)\cdot L^Y,$$

where $c \equiv (1-\alpha-\beta)^2 / \beta > 0$. A common market clearing wage of unskilled labor is obtained if and only if $\mu \leq 1$ and $\eta = 0$, according to lemma 3. Using (4.B.2) and (4.B.3) and the labor market clearing condition $L^X + L^Y = L$, this market clearing wage is given by

(4.B.4) $$w^{L*} = Ad\left(H^Y/L\right)^{\alpha},$$

where $d \equiv \beta^{\beta}(1-\alpha-\beta)^{2(1-\alpha-\beta)}(\beta+(1-\alpha-\beta)^2)^{\alpha} > b > 0$. Rewriting $w^{LY} = \eta w^{H*} + (1-\eta)w^{L*}$ from lemma 4.2 yields

(4.B.5) $$\frac{w^{LY}}{w^{H*}}\left(1-(1-\eta)\frac{w^{L*}}{w^{LY}}\right) - \eta = 0.$$

Substituting (4.19), (4.B.2) and (4.B.4) into (4.B.5), one gets

(4.B.6) $$\frac{\beta H^Y}{\alpha L^Y}\left(1-(1-\eta)\mu^{1-\alpha-\beta}\left(\frac{\beta+(1-\alpha-\beta)^2}{\beta}\frac{L^Y}{L}\right)^{\alpha}\right) - \eta = 0.$$

(4.B.6) defines L^Y implicitly as function of $L, \eta, \mu, \alpha, \beta$ and H^Y (where the latter is endogenously determined). Applying the implicit function theorem to (4.B.6) yields

(4.B.7) $$\frac{\partial L^Y}{\partial H^Y}\frac{H^Y}{L^Y} \in (0,1) \text{ if and only if } 0 < \eta < 1,$$

(4.B.8) $$\frac{\partial L^Y}{\partial H^Y}\frac{H^Y}{L^Y} = 1 \text{ if and only if } \eta = 1,$$

which proves lemma 4.4 (ii). Part (i) follows from (4.B.7) together with (4.19). Finally, part (iii) follows from (4.B.3) and part (ii). □

Appendix 4.C. Transitional dynamics

In this appendix, the transition path towards the balanced growth equilibrium is formally derived. Using (4.19) and (4.B.2), the skilled wage rate after t innovations in terms of aggregate productivity equals

$$\omega_t^{H*} = \alpha(d/\mu)^{1-\alpha-\beta}\left(L_t^Y / H_t^Y\right)^{1-\alpha}. \tag{4.C.1}$$

Rewriting (4.7), the expression for the marginal productivity of skilled labor in research, by using (4.2), (4.6), (4.8) and lemma 4.1 one obtains (after dividing by A_t)

$$\omega_t^{HR} = f'(H_t^R)\frac{\gamma\left(\dfrac{a(H_{t+1}^Y)^\alpha (L_{t+1}^Y)^\beta}{(\omega_{t+1}^{LX})^{1-\alpha-\beta}}\right)^{\frac{1}{\alpha+\beta}}}{r+f(H_{t+1}^R)}, \tag{4.C.2}$$

Substituting from(4.B.2) into (4.C.2) yields

$$\omega_t^{HR} = f'(H_t^R)\frac{(\alpha+\beta)(1-\alpha-\beta)(d/\mu)^{1-\alpha-\beta}\gamma(H_{t+1}^Y)^\alpha (L_{t+1}^Y)^{1-\alpha}}{r+f(H_{t+1}^R)}. \tag{4.C.3}$$

In equilibrium the right hand sides of (4.C.1) and (4.C.3) must be equal if a positive amount of labor is used in research, according to lemma 4.2. Hence, if $H_t^Y = H - H_t^R < H$ for all $t \geq 0$, a perfect foresight equilibrium $(H_0^Y, H_1^Y, H_2^Y, \ldots)$ satisfies

$$A(H_t^Y) = B(H_{t+1}^Y), \text{ where} \tag{4.C.4}$$

$$A(H_t^Y) \equiv \frac{\alpha(d/\mu)^{1-\alpha-\beta}\left(L_t^Y / H_t^Y\right)^{1-\alpha}}{f'(H-H_t^Y)}, \tag{4.C.5}$$

$$B(H_{t+1}^Y) \equiv \frac{(\alpha+\beta)(1-\alpha-\beta)(d/\mu)^{1-\alpha-\beta}\gamma(H_{t+1}^Y)^\alpha (L_{t+1}^Y)^{1-\alpha}}{r+f(H-H_{t+1}^Y)}. \tag{4.C.6}$$

Using (4.B.7) and (4.B.8), we obtain $A'(H_t^Y) \leq 0$ and $B'(H_{t+1}^Y) > 0$, respectively. Hence, there is a non-positive relation concerning the use of skilled labor in production between two subsequent periods t and $t+1$; that is

$$\frac{dH_{t+1}^Y}{dH_t^Y} = -\frac{A'(H_t^Y)}{B'(H_{t+1}^Y)} \leq 0, \tag{4.C.7}$$

where the latter holds with equality if and only if $\eta = 1$ and $f(\cdot)$ is a linear function in H^R. Fig. 4.1 shows a transition path of the economy converging towards its steady state $\hat{H}^Y$ (simultaneously determining unskilled employment and wages) as a clockwise spiral starting from H_0^Y.

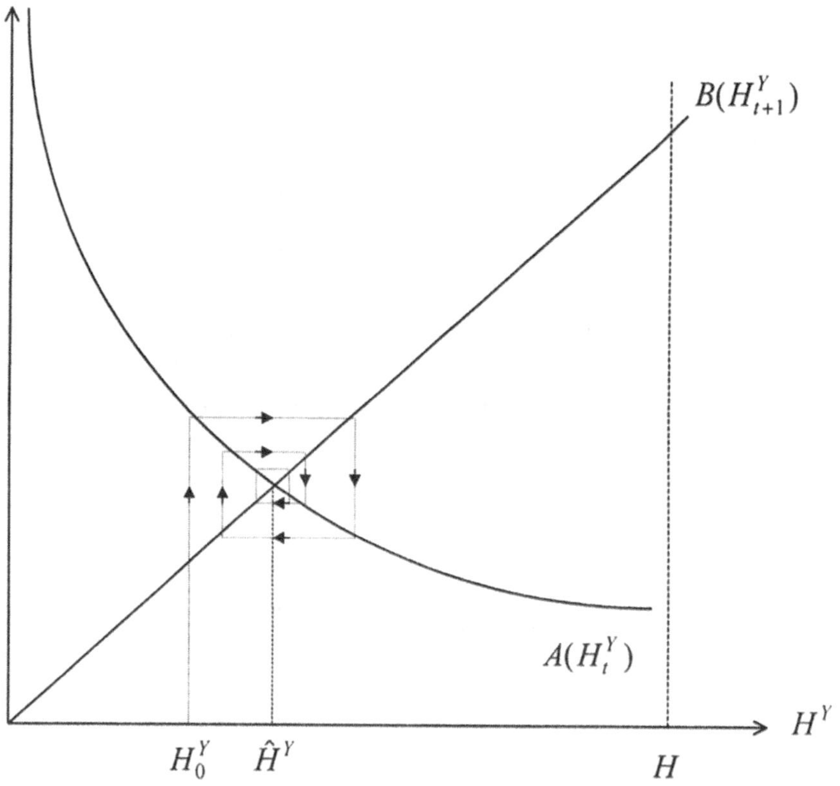

Figure 4.1: Transitional dynamics towards the steady state equilibrium if $\eta < 1$ or $f''(\cdot) < 0$.

Appendix 4.D. Allocation of the skilled labor force in steady state

A sufficient condition for $\hat{H}^R > 0$ is $A(H) < B(H)$ which is equivalent to $\gamma\alpha^{-1}(\alpha+\beta)(1-\alpha-\beta)f'(0)H > r$. If a positive amount of labor is used in research, then the steady state value of skilled labor in production is given by

(4.D.1) $$A(\hat{H}^Y) = B(\hat{H}^Y),$$

using (4.C.5), (4.C.6) and the definition of a stationary (balanced growth) equilibrium. (4.D.1) is equivalent to

(4.D.2)
$$\gamma(\alpha+\beta)\alpha^{-1}(1-\alpha-\beta)f'(H-\hat{H}^Y)\hat{H}^Y - \left(r + f(H-\hat{H}^Y)\right) = 0.$$

Applying the implicit function theorem to (4.D.2) yields

(4.D.3) $$\frac{\partial \hat{H}^Y}{\partial r} > 0\,, \quad \frac{\partial \hat{H}^Y}{\partial H} = 1 - \frac{\partial \hat{H}^R}{\partial H} \in (0,1)\,, \quad \frac{\partial \hat{H}^Y}{\partial \gamma} < 0$$

as stated in section 4.4. □

Appendix 4.E. Proof of propositions 4.1-4.3

Comparative static results with respect to the parameters r, γ and H directly follow from (4.D.3) and lemma 4.4. Note that the parameters η, μ and L do not affect the allocation of skilled labor, according to (4.D.2). Results with respect to η, μ and L thus follow from differentiating L^Y, which is implicitly defined as function of these parameters in (4.B.6), (4.B.3) and (4.19). □

Chapter 5. Where do we go from here? Conclusions and open questions

This book has examined the relationship between inequality, growth and technological change along two lines: first, the impact of wealth/income inequality on an economy's growth rate (of output, capital, consumption and productivity), and, second, the impact of both technological change and capital accumulation on the income distribution. Concerning the latter line, the book has followed the recent literature in focusing primarily on the impact of technological change on earnings inequality.

On the one hand, the book has been an attempt to provide a comprehensive picture about these relationships that have long played an important role (and still does) for both macroeconomic theory and economic policy. On the other hand, it has motivated and analyzed new issues that have emerged from the discussion. The new aspects concern the role of individual demand for public (redistribution) policy, the role of equity considerations of individuals and the role of non-production activities in an economy for the relationship between inequality and growth. In line with the recent literature on 'inequality, economic growth and technological change', the analysis has been based on endogenous growth models with heterogeneous individuals.

This final chapter summarizes the results of this study in the light of the 'old debate' about the relationship between inequality and growth, discusses policy implications and outlines both open questions and further research possibilities.

What can we conclude regarding the impact of inequality on growth?

According to a standard view, (high) inequality is needed to provide incentives to put effort in risky projects (e.g. if there is a random return to secondary schooling).[1] Moreover, if the marginal propensity to save is non-decreasing in income, total savings and thus growth may increase with the level of inequality. Public redistribution measures, however, are viewed as depressing savings and investment by lowering the return to capital. These views suggest that policy makers face a trade-off between redistribution and the growth rate of per capita income. If this trade-off is resolved suboptimally in the sense that too much growth is sacrificed from a social

[1] Aghion and Bolton (1997) have shown that this result can be overturned if credit markets are imperfect. For the role of imperfect credit markets for inequality and growth, see chapter 1 (section 1.4).

point of view, redistribution should be lowered in order to raise inequality of net income and thus investment incentives.[2]

However, in the beginning of the 1990s, the view that high inequality fosters growth has been seriously challenged by the empirical finding in cross-country studies that, if anything, income inequality is negatively related to both the ratio of total investment in physical capital to national income and the growth rate of per capita income. This finding has stimulated a lot of theoretical research.

According to the so-called 'politico-economic approach', high inequality leads to high redistribution levels through the political process. In turn, if redistribution is financed by taxation of income from accumulated factors like human or physical capital, both capital accumulation and growth of per capita income slow down. One should note that this approach is *not* at odds with the view that inequality of *net* income is good for growth for incentive reasons. This is because, in line with the cross-country evidence, politico-economic growth models address the link between inequality of *gross* income and growth. However, to date, there is no evidence supporting the politico-economic approach. On the one hand, taxation may indeed negatively affect growth. However, on the other hand, inequality of gross income even seems to have an adverse effect on redistribution and taxation, respectively, according to cross-country regression analysis. Other explanations of the finding that inequality is negatively related to growth are

---

[2] One should note that many models in which a trade-off between redistribution and growth exists imply that *any* redistribution policy financed by growth-depressing taxation is *constrained efficient* in the sense that a social planner with the same redistribution policy in hand could not raise efficiency. (See also the discussion of welfare effects of taxation in chapters 2 and 3). From the set of constraint efficient allocations a 'second-best' utility possibility set can be derived. Thus, the socially optimal solution to the trade-off between redistribution and growth depends on the social welfare function one wishes to employ [e.g. Grossmann (1999b)]. However, both policy makers and economic policy advisors often seem to believe that higher growth would unambiguously raise welfare. On the one hand, this may be due to a misinterpretation of the impact of taxation in standard representative agent growth models. (See e.g. Lucas (1990) and Rebelo (1991) for an analysis of taxation effects in representative agent endogenous growth models). However, if all individuals would be identical (so that redistribution is meaningless) and the tax revenue would not be used for publicly provided goods and services, it is trivial that cutting growth-depressing taxation would raise efficiency and thus would raise welfare unambiguously. (See e.g. Putterman, Roemer and Silvestre (1998) for this point). On the other hand, many policy makers seemingly try to maximize rather 'conservative' social welfare functions.

better supported by the data. According to the so-called 'capital market imperfections approach', when inequality is high, more people need to borrow in order to finance education (all other things equal). However, individuals may be credit-constrained if capital markets are imperfect. The main result is that growth may be slow when inequality is high due to an adverse effect of inequality on human capital accumulation.[3] This suggest that both income redistribution towards the poor and educational subsidies would be growth-promoting. High inequality may also be associated to both sociopolitical instability and high fertility rates. According to the former approach, sociopolitical instability contributes to an insecurity of property rights which in turn dampens investment incentives. According to the latter approach, high birth rates imply that parents do not invest much in the education of their children, which is also bad for growth. The evidence for both channels, although somewhat mixed, is rather supportive.

In sum, recent research suggests that there does not exist a trade-off between redistribution and economic growth in a cross-section of countries. However, it is not possible to conclude from cross-country evidence that redistribution policy would unambiguously spur growth in, say, Western Europe, the U.S. or Japan. This is because the empirically supported channels through which redistribution positively affects growth are likely to play a role primarily in less developed countries (which are over-represented in the outlined cross-country studies). For instance, Western democracies seem to be politically stable and fertility rates are rather low, irrespective of the level of inequality.[4] Moreover, in developed countries secondary schooling is encouraged by public education institutions and/or educational subsidies. In addition, unlike in developing countries where children are often supporting their families, in developed countries opportunity costs of sending children to school are very low on average. Thus, in these countries, capital market imperfections may not play a crucial role for the total investment in education. But most importantly, controlling for 'productive' public expenditure (like investment in infrastructure and education) taxation

---

[3] Two different kinds of models can generate this result. First, an individual schooling investment may be indivisible, so that the number of individuals who (are able to) finance this fixed capital outlay determines the total amount of investment. Second, if marginal individual returns to schooling investments are decreasing, the distribution of schooling investments among individuals is crucial for the aggregate amount of human capital.

[4] However, there is evidence that in the U.S. earning opportunities of low-skilled workers are adversely related to crime rates [e.g. Freeman (1996)]. This kind of negative externality of higher inequality or less redistribution, respectively, is in fact consistent with the sociopolitical instability approach even for otherwise stable democracies.

in developed countries is found to slow down growth.[5] Thus, although the politico-economic approach has been falsified, empirical evidence is not inconsistent with the view that developed countries are willing to trade-off some growth for more equality.

Chapter 2 has provided a new politico-economic explanation why this may be the case.[6] It has been argued that individual demands for redistribution can be attributed to *status concerns* of individuals. Redistribution is assumed to be financed by growth-depressing taxation.

Chapter 2 has started out with the observation that in the developed world per capita income does not seem to be positively related to average happiness or satisfaction levels of individuals, respectively. This raises doubts about positive welfare effects of growth. It has been shown that negative consumption externalities are consistent with this kind of evidence. That is, all other things equal, the consumption levels of others negatively affect individual well-being. The growth model in chapter 2 has specified this idea in assuming that individuals are concerned about their relative consumption level. If relative consumption (or, more generally, relative standing) determines one's social status, status is a good which is in fixed supply even in a growing economy. The basic idea of the model is that status-concerned individuals with less than average (human capital) income can gain status if income is redistributed. In other words, given a framework in which redistribution depresses growth, poorer individuals are willing to trade-off growth for status. (In contrast, richer individuals do not face such a trade-off between status and growth). Thus, if the pivotal voter in a standard voting framework is relatively (capital-) poor, status concerns negatively affect growth through the political process. (The pivotal voter does not have to be median income earner).

Of course, majority voting models may not be a good representation of the political process, and thus may be unsatisfactory as a positive theory of redistribution (which has long been discussed elsewhere). So one might want to focus on the individual demand for redistribution rather than looking at the voting equilibrium. Doing so, from a *normative* point of view one can derive the following policy implication. Because in the presence of status concerns redistribution is less costly (in terms of sacrificing growth for equality) for poorer individuals than standard models predict, the socially optimal level of redistribution may be higher than standard models imply (i.e. if the socially optimal level of redistribution is derived from a social

---

[5] Kneller, Bleaney and Gemmell (1999). See also chapter 1 (section 1.4) for further discussion.

[6] However, the analysis could not contribute to understand recent *changes* in redistribution policies in developed countries; see also the discussion below.

welfare function that puts sufficiently high weight on relatively poor individuals).

Moreover, according to a popular view, negative consumption externalities imply that savings and thus investment-driven growth are depressed also apart from the politico-economic channel (i.e. for a *given* redistribution policy). According to this argument, people spend too much in order to 'keep up with the Joneses'. However, in an intertemporal model with relative consumption preferences, there is also the effect that savings allow individuals to increase their status in the future. Thus, regarding economic policy, one should not necessarily design policies (like taxation of consumption) to internalize negative consumption externalities, as has previously been suggested.[7]

Chapter 3 has dealt with the failure of the politico-economic approach to explain a negative relationship of inequality and growth. That is, it has been shown in chapter 3 that the level of redistribution or taxation, respectively, does not necessarily increase with inequality of gross income in voting equilibrium. The basic idea in chapter 3 has been that, unlike in standard politico-economic growth models, the tax revenue is used for the public provision of (public or private) goods and services rather than for redistributive transfers. Nevertheless, taxation is redistributive in the model.[8] That is, taxes are levied in a non-lump sum fashion whereas the tax revenue is redistributed lump sum through publicly provided goods and services. As in chapter 2, majority voting and growth-depressing taxation is considered. However, whereas in chapter 2 a trade-off between redistributive transfers and growth has been examined, in chapter 3 the trade-off is between the level of public consumption and growth.

In chapter 3 it has been shown that, contrary to the conventional wisdom, richer individuals may demand *higher* income taxes in order to finance government consumption. The reason is the following. An increase in individual income induces an income and a substitution effect. The income effect simply implies that richer individuals demand higher levels of the publicly provided good (i.e. public consumption is a normal good for individuals in the model). The substitution effect is due to the assumption that the individual tax share (i.e. the contribution share to the publicly provided good) increases with individual income. Thus, due to the redistributive tax system, the substitution effect of an increase in individual income works in the opposite direction than the income effect. It has been shown that the income effect is even more likely to dominate in an

[7] E.g. Frank (1985a,b).

[8] Concluding that taxation reduces inequality is also in line with the relevant empirical studies on inequality, taxation and growth [e.g. Perotti (1996)].

intertemporal than in a static context. In both the static model and the growth model, higher public consumption comes at the cost of lower private consumption. In addition, in the growth model, an increase in the level of public consumption also reduces the rate of growth due to higher taxation. According to the latter kind of trade-off, taxation has an impact on the growth rate of individual income which is similar for all individuals. However, the impact of taxation on private consumption levels substantially differs among individuals. The reason is the following. In the growth model, richer individuals endogenously have higher saving rates. If rich individuals reduce savings after an increase in taxes, they may maintain their current consumption levels. But the reduction in overall savings slows down growth, thus affecting poor individuals as well. In contrast, poor individuals are stuck with low savings to begin with and thus reduce their private consumption levels if tax rates increase. Thus, their trade-off between private and public consumption is more pronounced than that of rich individuals. As a result, poor individuals may demand low taxes.

In sum, chapter 3 has shown that standard results about the relationship of inequality and public expenditure in the context of growth may be overturned even when maintaining the standard voting framework. In addition to this theoretical point, chapter 3 also provides empirical evidence for the derived hypothesis that income inequality in a country (measured by the Gini coefficient) is not positively related to the public consumption share (i.e. total public consumption expenditure divided by GDP) in both OECD countries and democracies.

This leads us to open questions concerning the relationship between inequality, redistributive (i.e. non-productive) public expenditure and growth. First, the existing literature (including chapters 2 and 3) cannot explain differences in the levels of redistribution and growth among countries. For instance, why is there little redistribution in the U.S. and the UK but a comparably high level of redistribution in Scandinavian countries and Germany? In addition, one may ask whether or not low redistribution levels in the U.S. have contributed to the recent surge in the U.S. growth rate. And, as the book has emphasized the role of equity concerns on an *individual* level, the role of *social organizations* and *cultural factors* for both redistribution and growth should be studied in more detail.[9] Second, why do policy makers in developed countries seem to be increasingly keen on cutting back redistribution levels? For example, taxation of both capital income and wealth has recently been lowered almost everywhere in the developed world, but a convincing positive theory for these rather dramatic developments has not been provided yet.

---

[9] See Corneo and Jeanne (1998, 1999a) for first attempts.

Besides the impact of personal income/wealth inequality on growth, this book has also focused on the impact of both technology-driven and investment-driven growth on the functional and personal income distribution, respectively. Especially, the question how technological change affects the distribution of earnings has been examined. This has recently been discussed under the label of *skill-biased technological change*, a discussion that has been stimulated by a large body evidence for a substantial shift of (relative) labor demand against low-skilled workers.

In contrast to this current debate, classical, neoclassical, Marxian and Keynsian distribution and growth theory, respectively, was concerned with the determinants and the evolution of the *functional* income distribution. As briefly reviewed in chapter 1, whereas David Ricardo and Karl Marx hypothesized substantial changes of the functional income distribution over time, subsequent theories implied that this distribution would be rather stable.[10] In fact, changes in the functional income distribution in developed countries have long been minor. However, as discussed in chapter 1, there is evidence for a substantial decline in the labor share of national income in the 1980s in many developed countries. Although there is still some debate about how this share is to measure, this view is also supported by increases in real interest rates in the 1980s.

The literature on the evolution of *personal* income inequality over time has been pioneered by Simon Kuznets (1955), who hypothesized an inverted U-shaped relationship of personal income inequality and national income per capita. In chapter 1, new evidence *against* the Kuznets hypothesis in a *cross-section* of countries has been presented. However, in order to test the original Kuznets hypothesis one has to look at the relationship between inequality and per capita income within countries *over time*. (Up to date this kind of analysis is severely restricted by the lack of time series income inequality data with decent quality).

But at least the Anglo-American experience of rising earnings inequality across and within observably different groups of workers in the last two decades allows to draw conclusions against the original Kuznets hypothesis. Provided that the earnings distribution is an important determinant of the overall income distribution (which is a plausible and empirically supported hypothesis), there is no reason to believe that developed countries have reached a point where income inequality is declining (or at least non-increasing) over time.

---

[10] As outlined in chapter 1 (section 1.1), also Ricardo saw the economy approaching to a steady state functional income distribution, but one could interpret the transition dynamics he had in mind as being characterized by a lower 'speed of convergence' than, say, in neoclassical growth models.

The supply of high-skilled labor (relative to low-skilled labor) has substantially increased in developed countries. One might have expected this to cause a decline in wage dispersion across skill groups. However, in the U.S. and the UK, wage dispersion across skill groups has even surged, and, at least, it has not declined elsewhere. The widely accepted explanation for this fact is that technological change has led to *skill-biased* shifts in the production function, implying a shift in relative labor demand in favor of better qualified workers.[11]

However, as has been criticized in this book, the hypothesis of skill-biased technological change as determinant for rising (or non-decreasing) wage inequality is almost tautological.[12] In fact, the skill-bias hypothesis raises far more questions than it answers. Technological changes may indeed substantially affect the dispersion of earnings. But although very sophisticated models of endogenous skill-biased technological change have been proposed, this literature has left the impact of recent technological changes on the production process and thus on both labor and goods markets almost entirely in the dark.

According to the empirical skill-bias literature, relative labor demand shifts have gone along with a wide-spread computerization and adoption of new information and communication technologies. Moreover, one has observed substantial skill-upgrading, i.e. an increase in the employment share of skilled workers, mainly within firms and industries. In most empirical studies, skilled employment has been approximated by non-production employment (due to better data availability). In fact, non-production activities mainly encompass white-collar work, and white-collar workers are predominantly skilled. However, given the large increase in the supply of skilled labor it is not really surprising that the employment share of better qualified workers has surged. Of course, the challenging question from a theoretical point of view is why at the same time wage dispersion did not decline. However, the only answer of the skill-bias literature is that 'new technologies complement skills'.[13] But, taken literally, this interpretation

---

[11] If one accepts the view that relative labor demand shifts are technology-related, one may generally conclude that it is the unpredictability of the nature of technological change which historically is likely to falsify the Kuznets hypothesis. In fact, Kuznets (1955) himself did not discuss the relationship between the personal income distribution and technological change at all (see chapter 1, section 1.3).

[12] As Thurow (1998) polemically points out, "the only evidence for such a technical shift is the rising wage gap itself" (p. 27).

[13] This phrase is contained in the title of a prominent paper by Acemoglu (1998), who provides an explanation for this kind of relationship.

suggests that especially computer users are the winners from the 'computer revolution'. However, this does not seem to be the case.[14]

Chapter 4 has suggested an alternative explanation for the fact that wage inequality across skill groups has not declined despite the marked increase in the relative supply of qualified workers. In contrast to the theoretical skill-bias literature, production and non-production activities have been analytically distinguished. It has been argued that, in accordance with the empirical evidence, technological change has caused resources to shift towards non-production activities. For instance, one may imagine that in order to reap the benefits from new information and communication technologies, firms have to reorganize work. According to the model, this restructuring process raises productivity in a Hicks-neutral rather than in a skill-biased way. Nevertheless, relative marginal productivity of high-skilled workers increases, but only because of the reallocation of skilled labor towards non-production. This is because such an employment shift reduces the marginal productivity of low-skilled labor in production, if high-skilled and low-skilled labor in production are technological complements.

In the model, non-production work has been specified as R&D, but is supposed to reflect any non-production activity that raises the productivity or revenue of firms. This also includes marketing activities, as well as training, coordination and supervision of workers. Non-production employment shifts in the model may reflect reorganization within firms or 'outsourcing' of business services, respectively.

Moreover, the model of chapter 4 has taken into account the empirical evidence about the role of equity concerns for the distribution of earnings. That is, the impact of fairness considerations on earnings inequality has been analyzed. In the model, workers not only socially compare themselves with workers of other skill groups, like in Akerlof and Yellen (1988, 1990), but also within the own skill group. Firms observe that workers reduce their effort levels when they feel treated unfairly.[15] Due to social comparisons of low-skilled with high-skilled workers, wage inequality across skill groups is reduced by efficiency wage payments. That is, relative wages are sticky and low-skilled jobs are rationed. In the model of Akerlof and Yellen (1990), in which one representative firm produces a homogenous consumption good, this necessarily implies unemployment of low-skilled workers. In contrast, in our model low-skilled workers may also find employment elsewhere in the economy, i.e. in intermediate goods production. (In non-production, only high-skilled workers can be employed). Thus, whether or not sticky relative wages in final goods production leads to aggregate unemployment depends

[14] DiNardo and Pischke (1997).

[15] Akerlof and Yellen (1990).

on the flexibility of wages in intermediate goods production. Since low-skilled workers in intermediate goods production make social comparisons with low-workers in final goods production, efficiency wage considerations of firms lead to rigidity of absolute wages as well. In the model, this absolute wage stickiness is necessary for unemployment of the low-skilled.

If skilled workers are shifted towards non-production activities, changes in both wage dispersion and employment of the low-skilled depend on social comparisons across and within skill groups, respectively. If no social comparisons would be made at all, only wage inequality across skill groups would change. In the presence of fairness considerations across skill groups in final goods production, a non-production employment shift in this industry induces wage pressure in intermediate goods production. If no within-group social comparisons would be made, this would lead to within-group wage inequality. But if there are also within-group social comparisons (across industries), the labor reallocation towards non-production raises unemployment rather than within-group wage inequality.

In sum, chapter 4 has shown that by looking at both technology-related restructuring processes and equity concerns of workers one can gain additional insights and thus a more differentiated picture about the consequences of technological change for the earnings distribution.

Future research should thus study the impact of technological change on work organization and labor market outcomes in much more detail. There is no doubt that firm organizations, the structure of industries and patterns of working life have recently changed in a substantial way.[16] For instance, firms have adopted organizational environments which are characterized by greater emphasis on team work, information sharing, more job complexity, job rotation, decentralized structures, and so on. These developments are likely to raise demand for workers with good social and communication skills, the ability to perform multiple tasks and the ability to accommodate to changes in organizational structures. These features are barely reflected in existing macroeconomic models. Perhaps most importantly, macroeconomic theory neglects the fact that resources have to be expended to adopt and maintain new organizational structures which are supposed to raise productivity. Although the analysis in chapter 4 has been rather rudimentary with respect to these issues, it has demonstrated how views about labor market phenomena, and thus about the relationship of technological change and earnings inequality, can change by taking the latter fact into account.

---

[16] For a recent survey and evidence of changes in the organization of work, see OECD (1999c). See also Lindbeck and Snower (1996), Kremer and Maskin (1996) and the discussion in Aghion, Caroli and García-Peñalosa (1999) for first attempts to relate the reorganization of firms to wage inequality.

Moreover, if social and communication skills become more important, wage inequality within education groups may surge, since formal education is only imperfectly related to these skills. Equity concerns, as analyzed in this book, may put boundaries to such changes in the wage distribution, with possibly interesting macroeconomic effects.

Technological change and reorganization of work may also be related to developments currently discussed under the term 'globalization'. For instance, the wide-spread adoption of information technologies may have cut costs of both coordination of decentralized production processes and direct communication of workers employed at different locations of multinational enterprises.[17] Thus, decreasing information and communication costs may allow firms to profitably extend their markets. Technological change may also lead to economies of scale and market power which would be consistent with the recent wave of firm mergers.[18] Another issue which has largely been ignored in the mainstream economic literature concerns 'outsourcing' of business services by firms.[19] The waves of both firms mergers and outsourcing also stress the importance of looking into the organization of production for labor market phenomena like the evolution of earnings inequality. Although discussed extensively in the business literature, the macroeconomic impact of these dramatic changes is largely unexplored yet.

Other open questions concern the very different patterns of wage distribution and unemployment across countries. For instance, whereas in the U.S. and the UK wage inequality has surged, it has been fairly stable almost everywhere else. The widely-accepted hypothesis that in Europe strong labor market institutions have prevented a rise in wage inequality at costs of an increase in unemployment is not convincing. First, the Anglo-American employment record of low-skilled workers is, if anything, worse than in continental Europe.[20] Second, in contrast to the U.S. and the UK, in continental Europe also wages at the top of the wage distribution seem to be stable (relative to the median). But it is unlikely that labor market

[17] See Falkinger (2000, ch. 4) for a discussion of employment consequences of such 'globalization' characteristics.

[18] As indicated in chapter 1 (section 1.1), Karl Marx suggested adverse effects of firm concentration processes for the situation of workers.

[19] See Burda and Dluhosch (1998) for an attempt to explain these outsourcing activities and analyze their labor market effects.

[20] As pointed out by Nickell and Bell (1997, p. 309), "there seems to be no evidence that the unemployment rate effects [of the fall in relative demand for unskilled workers] are any more severe in countries where the wage effects are minimal or perverse". See Falkinger and Grossmann (1999) for both further discussion and a model addressing this issue.

institutions prevent wages from *rising.*[21] As outlined in chapter 1, the prominent theoretical literature of the labor market impact of technological change has mainly focused on the Anglo-American experience. However, its difference to the European experience raises doubts on the generality of these approaches. Of course, institutions may play an important role for labor market effects of technological change, and increases in aggregate unemployment figures in Europe are striking. But which role institutions play for both the *structure* of unemployment (e.g. unemployment of different education, occupation or age groups) and the wage distribution is not well understood yet.

Finally, so far the newer literature on changes in income inequality has mainly focused on earnings inequality (including this book). But in the last, say, two decades there has been a departure from what Nicolas Kaldor (1956) has called the "historical constancy" of the functional income distribution. In fact, Kaldor (1956) disagreed with Ricardo (1821) who stated in his preface that "in different stages of society the proportions of the whole produce which will be allotted to [...] rent, profits and wages will be essentially different", concluding that one "would almost be inclined to say the opposite - that in different stages of (capitalist) society the proportions of national income allotted to wages, profits, etc., are *essentially similar*" [Kaldor (1956, p. 83; italics original)]. Moreover, he proceeded with the statement that "no hypothesis as regards the forces determining distributive shares could be intellectually satisfying unless it succeeds in accounting for the relative *stability* of these shares in the advanced capitalist economies" [Kaldor (1956, p. 83-84; italics added)]. Today, it rather seems that we lack an explanation for recent *declines* of the labor share in national income in advanced countries. Moreover, as pointed out by Atkinson (2000), also the impact of changes in the "distributive shares" for the personal income distribution has to be explored.

After over two centuries of research in the field of inequality and growth, the unpredictability of changes in production technologies, social organizations and political structures has caused and probably will always cause theoretical explanations to lag behind the changes of growth and inequality patterns. These are certainly good prospects for economists who are interested in this field.

---

[21] See Atkinson (2000) for this argument.

List of figures and tables

References

Abel, A. (1990). Asset prices under habit formation and catching up with the Joneses, *American Economic Review* 40, 38-42.

Abraham, K. G. and S. N. Houseman (1995). Earnings inequality in Germany, in: Freeman, R. B. and L. F. Katz (eds.), Differences and changes in wage structures (Chicago: University of Chicago Press).

Abraham, K. G., J. R. Spletzer and J. C. Stewart (1999). Why do different wage series tell different stories?, *American Economic Review Papers and Proceedings* 89, 34-39.

Acemoglu, D. (1998). Why do new technologies complement skills? Directed technical change and wage inequality, *Quarterly Journal of Economics* 113, 1055-1089.

Acemoglu, D. (1999). Changes in unemployment and wage inequality: an alternative theory and some evidence, *American Economic Review* 89, 1259-1278.

Adelman, I. (1961). Theories of economic growth and development (Stanford: Stanford University Press).

Agell, J. and P. Lundborg (1995). Theories of pay and unemployment: survey evidence from Swedish manufacturing firms, *Scandinavian Journal of Economics* 97, 295-307.

Agénor, P. and J. Aizenman (1997). Technological change, relative wages, and unemployment, *European Economic Review* 41, 187-205.

Aghion, P. and P. Bolton (1997). A theory of trickle-down growth and development, *Review of Economic Studies* 64, 151-172.

Aghion, P., E. Caroli and C. García-Peñalosa (1999). Inequality and economic growth: the perspective of new growth theories, *Journal of Economic Literature* 37, 1615-1660.

Aghion, P. and P. Howitt (1992). A model of growth through creative destruction, *Econometrica* 60, 323-352.

Aghion, P. and P. Howitt (1994). Growth and unemployment, *Review of Economic Studies* 61, 477-494.

Aghion, P and P. Howitt (1998). Endogenous growth theory (Cambridge: MIT Press).

Ahn, S. (1999). Technology upgrading with learning costs: a solution for two 'productivity puzzles', OECD Economic Department Working Paper No. 220.

Ahluwalia, M. S. (1976). Inequality, poverty and development, *Journal of Development Economics* 3, 307-342.

Aitchison, J. and J. A. Brown (1966). The lognormal distribution with special reference to its uses in economics (Cambridge: Cambridge University Press).

Akerlof, G. A. (1982). Labor contracts as partial gift exchange, *Quarterly Journal of Economics* 82, 543-569.

Akerlof, G. A. and J. Yellen (1988). Fairness and unemployment, *American Economic Review Papers and Proceedings* 78, 44-49.

Akerlof, G. A. and J. Yellen (1990). The fair wage-effort hypothesis and unemployment, *Quarterly Journal of Economics* 105, 255-283.

Alesina, A. and R. Perotti (1996). Income distribution, political instability, and investment, *European Economic Review* 40, 1203-1228.

Alesina, A. and D. Rodrik (1994). Distributive politics and economic growth, *Quarterly Journal of Economics* 109, 465-490.

Arrow, K. J. (1962). The economic implications of learning by doing, *Review of Economic Studies* 29, 155-173.

Atkinson, A. B. (1997). Bringing income distribution in from the cold, *Economic Journal* 107, 297-321.

Atkinson, A. B. (2000). The changing distribution of income: evidence and explanations, *German Economic Review* 1, 3-18.

Atkinson, A. B. and J. E. Stiglitz (1980). Lectures on public economics (London: McGraw-Hill).

Autor, D. H., L. F. Katz and A. B. Krueger (1998). Computing inequality: have computers changed the labor market?, *Quarterly Journal of Economics* 113, 1169-1214.

Banerjee A. V. and A. F. Newman (1993). Occupational choice and the process of development, *Journal of Political Economy* 101, 274-298.

Barro R. J. (1990). Government spending in a simple model of economic growth, *Journal of Political Economy* 98, S103-S125.

Barro R. J. and X. Sala-i-Martin (1995). Economic growth (New York et al.: McGraw-Hill, Inc.).

Basett, W. F., J. P. Burkett and L. Putterman (1999). Income distribution, government transfers, and the problem of unequal influence, *European Journal of Political Economy* 15, 207-228.

Bean, C. and C. A. Pissarides (1993). Unemployment, consumption and growth, *European Economic Review* 37, 837-859.

Becker, G. S. and R. J. Barro (1988). A reformulation of the economic theory of fertility, *Quarterly Journal of Economics* 103, 1-25.

Becker, G. S., K. M. Murphy and R. Tamura (1990). Human capital, fertility, and economic growth, *Journal of Economy* 98, S12-S37.

Beckman, S. R., J. Formby, W. J. Smith, and B. Zheng (1997). An experimental examination of Pareto efficiency, envy, malice and the sentiment of equality, Working Paper 9703, University of Colorado-Denver, Center for Research on Economic and Social Policy.

Beißinger, T. and J. Möller (1999). Unemployment: theoretical explanations, in: H. Wagner (ed.), Globalisation and unemployment (Berlin et al.: Springer).

Bénabou, R. (1996). Inequality and growth, *NBER Macroeconomics Annual*, 11-74.

Benhabib, J. and A. Rustichini (1996). Social conflict and growth, *Journal of Economic Growth* 1, 125-142.

Bergstrom, T. C. and R. P. Goodman (1973). Private demands for public goods, *American Economic Review* 63, 280-296.

Bergström, W. J. and E. E. Panas (1992). How robust is the capital-skill complementarity hypothesis?, *Review of Economics and Statistics* 74, 540-552.

Berman, E., J. Bound and Z. Grichilis (1994). Changes in the demand for skilled labor within U.S. manufacturing: evidence from the annual survey of manufactures, *Quarterly Journal of Economics* 109, 367-397.

Berman, E., J. Bound and S. Machin (1998). Implications of skill-biased technological change: international evidence, *Quarterly Journal of Economics* 113, 1245-1279.

Bertola, G. (1993). Factor shares and savings in endogenous growth, *American Economic Review* 83, 1184-1198.

Bertola, G. (1996). Factor shares in OLG models of growth, *European Economic Review* 40, 1541-1560.

Bewley, T. F. (2000). Why wages don't fall during a recession (Cambridge: Harvard University Press).

Billingsley, P. (1979). Probability and measure (New York: Wiley).

Blanchard, O. (1997). Is there a core of usable macroeconomics?, *American Economic Review Papers and Proceedings* 87, 244-246.

Blau, F. D. and L. M. Kahn (1996). International differences in male wage inequality: institutions versus market forces, *Journal of Political Economy* 104 (4), 791-837.

Blinder, A. S. (1988). The challenge of high unemployment, *American Economic Review Papers and Proceedings* 78, 1-15.

Boadway, R. W. and D. E. Wildasin (1984). Public sector economics, 2nd Edition (Boston: Little, Brown & Company).

Bohl, M. (1996). Some international evidence on Wagner's law, *Public Finance* 51, 185-200.

Borcherding, T. E. and R. T. Deacon (1972). The demand for services of non-federal governments, *American Economic Review* 62, 891-901.

Bosworth, B., G. Burtless and J. Sabelhaus (1991). The decline in saving: evidence from household surveys, Brooking Papers of Economic Activity 1, 183-241.

Bourguignon, F. (1981). Pareto superiority of unegalitarian equilibria in Stiglitz' model of wealth distribution with convex saving function, *Econometrica* 49, 1469-1475.

Bourguignon, F. (1990). Growth and inequality in the dual model of development: the role of demand factors, *Review of Economic Studies* 57, 215-228.

Bresnahan, T. F. (1999). Computerisation and wage dispersion: an analytical reinterpretation, *Economic Journal* 109, F390-F415.

Browning, M. and A. Lusardi (1996). Household saving: micro theory and micro facts, *Journal of Economic Literature* 34, 1797-1855.

Buchholz, W. (1994). Das Weniger-Ist-Mehr-Paradoxon, *Wirtschaftsethische Perspektiven* 2, 177-193.

Buchholz, W. and W. Peters (2000). The overprovision anomaly of private public good supply, University of Regensburg (mimeo)

Burda, M. C. and B. Dluhosch (1998). Globalization and Eurpean labour markets, CEPR, Discussion Paper No. 1992.

Campano, F. and D. Salvatore (1988). Economic development, income inequality and Kuznets' U-shaped hypothesis, *Journal of Policy Modeling* 10, 265-280.

Campbell, J. Y. and J. H. Cochrane (1995). By force of habit: a consumption-based explanation of aggregate stock market behavior, NBER, Working Paper No. 4995.

Card, D. and A. B. Krueger (1995). Myth and measurement: the new economics of the minimum wage (Princeton: Princeton University Press).

Card, D., F. Kramarz and T. Lemieux (1999). Changes in the relative structure of wages and employment: a comparison of the United States, Canada, and France, *Canadian Journal of Economics* 32, 843-877.

Carroll, C. D. and D. N. Weil (1994). Saving and growth: a reinterpretation, *Carnegie-Rochester Conference Series on Public Policy* 40, 133-192.

Carroll, C., D. J. Overland and D. N. Weil (1997). Comparison utility in a growth model, *Journal of Economic Growth* 2, 339-367.

Caselli, F. (1999). Technological revolutions, *American Economic Review* 89, 78-102.

Cass, D. (1965). Optimum growth in an aggregative model of capital accumulation, *Review of Economic Studies* 32, 233-240.

Clark, J. B. (1899). The distribution of wealth (London: Macmillan).

Clark, A. E. and A. J. Oswald (1996). Satisfaction and comparison income, *Journal of Public Economics* 61, 359-381.

Cobb, C. W. and P. H. Douglas (1928). A theory of production, *American Economic Review* 18 (Supplement).

Cohen, D., A. Lefranc and G. Saint-Paul (1997). French unemployment: a transatlantic perspective, *Economic Policy* 25, 267-291.

Cole, H. L., G. J. Mailath and A. Postlewaite (1992). Social norms, savings behavior, and growth, *Journal of Political Economy* 100, 1092-1125.

Corneo, G. and H. P. Grüner (1996). Social limits to redistribution, University of Bonn Discussion Paper A520.

Corneo, G. and O. Jeanne (1997). On relative wealth effects and the optimality of growth, *Economics Letters* 54, 87-92.

Corneo, G. and O. Jeanne (1998). Social organization, status, and savings behavior, *Journal of Public Economics* 70, 37-51.

Corneo, G. and O. Jeanne (1999a). Social organization in an endogenous growth model, *International Economic Review* 40, 711-725.

Corneo, G. and O. Jeanne (1999b). Pecuniary emulation, inequality and growth, *European Economic Review* 43, 1665-1678.

Dasgupta, P. S. (1993). An inquiry into well-being and destitution (Oxford: Clarendon Press).

Deininger, K. and L. Squire (1996). A new data set measuring income inequality, *World Bank Economic Review* 10, 565-591.

Deininger, K. and L. Squire (1998). New ways of looking at old issues: inequality and growth, *Journal of Development Economics* 57, 259-287.

De Vivo, G. (1987). Ricardo, David, in: J. Eatwell, M. Milgate and P. Newman (eds.), The new Palgrave: a dictionary of economics (London: Macmillan).

DiNardo, J. E. and J. Pischke (1997). The returns to computer use revisited: have pencils changed the wage structure too?, *Quarterly Journal of Economics* 112, 291-303.

Dixit, A. K. (1976). The theory of equilibrium growth (Oxford: Oxford University Press).

Domar, E. (1946). Capital expansion, rate of growth and unemployment, *Econometrica* 14, 137-147.

Duesenberry, J. S. (1949). Income, savings and the theory of consumer behavior (Cambridge: Harvard University Press).

Easterly, W. and S. Rebelo (1993). Fiscal policy and economic growth: an empirical investigation, *Journal of Monetary Economics* 32, 417-458.

Easterlin, R. A. (1974). Does economic growth improve the human lot? Some empirical evidence, in: P. A. David and M. W. Reder (eds.), Nations and households in economic growth: essays in honor of Moses Abramovitz (New York: Academic Press, Inc.).

Easterlin, R. A. (1995). Will raising the income of all increase the happiness of all?, *Journal of Economic Behavior and Organization* 27, 35-47.

Economist (2000). Up and up it goes, February 12th 2000, 51.

Edin, P. and B. Holmlund (1995). The Swedish wage structure: the rise and fall of solidarity wage policy?, in: Freeman, R. B. and L. F. Katz (eds.), Differences and changes in wage structures (Chicago: University of Chicago Press).

Epple, D. and R. E. Romano (1996). Public provision of private goods, *Journal of Political Economy* 104, 57-84.

European Commission (1999). The competitiveness of European industry: 1999 report, Luxembourg.

Falkinger, J. (1999). Social instability and redistribution of income, *European Journal of Political Economy* 15, 35-51.

Falkinger, J. (2000). Macroeconomic equilibrium and the organization of work: a new theory of employment, University of Regensburg (mimeo).

Falkinger, J. and V. Grossmann (1999). Skills, unemployment and the costs of organizing jobs, University of Regensburg, Discussion Paper No. 321.

Falkinger, J. and J. Zweimüller (2000). Learning for employment, innovating for growth, *Journal of Institutional and Theoretical Economics* (forthcoming).

Fehr, E. and S. Gächter (1998). Reciprocity and economics: the economic implications of *homo reciprocans*, *European Economic Review* 42, 845-859.

Fehr, E. and S. Gächter (1999). Cooperation and punishment in public goods experiments, *American Economic Review* (forthcoming).

Fehr, E., S. Gächter and G. Kirchsteiger (1996). Reciprocal fairness and noncompensating wage differentials, *Journal of Institutional and Theoretical Economics* 152, 608-640.

Fehr, E., G. Kirchsteiger and A. Riedl (1993). Does fairness prevent market clearing? An experimental investigation, *Quarterly Journal of Economics* 108, 437-459.

Fehr, E., G. Kirchsteiger and A. Riedl (1998). Gift exchange and reciprocity in experimental markets, *European Economic Review* 42, 1-34.

Fehr, E. and K. Schmidt (1999). A theory of fairness, competition and cooperation, *Quarterly Journal of Economics* 114, 817-868.

Fershtman, C., K. M. Murphy and Y. Weiss (1996). Social status, education, and growth, *Journal of Political Economy* 104, 108-132.

Fiaschi, D. (1999). Growth and inequality in an endogenous fiscal policy model with taxes on labor and capital, *European Journal of Political Economy* 15, 727-746.

Figini, P. (1999). Inequality and growth revisited, Trinity College, Trinity Economic Papers Series, Paper No. 99/2.

Fortin, N. M. and T. Lemieux (1997). Institutional changes and rising wage inequality: is there a linkage?, *Journal of Economic Perspectives* 11, 75-96.

Frank, R. H. (1985a). Choosing the right pond: human behavior and the quest for status (New York: Oxford University Press).

Frank, R. H. (1985b). The demand for unobservable and other nonpositional goods, *American Economic Review* 75, 101-116.

Franz W. (1999). The European unemployment problem, Paper prepared for the CESifo Symposium No. 1 in Munich.

Freeman, R. B. (1995). Are your wages set in Beijing?, *Journal of Economic Perspectives* 9, 15-32.

Freeman, R. B. (1996). Why do so many young American men commit crimes and what might we do about it, *Journal of Economic Perspectives* 10, 25-42.

Futagami, K. and A. Shibata (1998). Keeping one step ahead of the Joneses: status, the distribution of wealth, and long run growth, *Journal of Economic Behavior and Organization* 36, 109-126.

Galor, O. and O. Moav (1999). From physical to human capital accumulation: inequality in the process of development, Hebrew University (mimeo).

Galor, O. and D. Tsiddon (1997). Technological progress, mobility, and economic growth, *American Economic Review* 87, 363-382.

Galor, O. and J. Zeira (1993). Income distribution and macroeconomics, *Review of Economic Studies* 60, 35-52.

Gastil R. D. and followers (various years). Freedom in the world (Westport: Greenwood Press).

Gemmell, N. (1990). Wagner's law, relative prices, and the size of the public sector, *The Manchester School* 43, 361-377.

Giles, C., A. Gosling, F. Laisney and T. Geib (1998). The distribution of income and wages in the UK and West Gemany 1984-1992, Institute for Fiscal Studies, London.

Glomm, G. and B. Ravikumar (1992). Public versus private investment in human capital: endogenous growth and income inequality, *Journal of Political Economy* 100, 818-834.

Goldin, C. and L. F. Katz (1998). The origin of technology-skill complementary, *Quarterly Journal of Economics* 113, 693-732.

Gollop, F. M. and D. W. Jorgenson (1980). U.S. productivity growth by industry, in: Kendrick, J. W. and B. N. Vaccara (eds.), New developments in productivity measurement and analysis (Chicago: University Chicago Press).

Gottschalk, P. and T. M. Smeeding (1997). Cross-national comparisons of earnings and income inequality, *Journal of Economic Literature* 105, 633-687.

Gregg, P. and A. Manning (1997). Skill-biassed change, unemployment and wage inequality, *European Economic Review* 41, 1173-1200.

Grichilis, Z. (1994). Productivity, R&D, and the data constraint, *American Economic Review* 84, 1-23.

Grossman, G. and E. Helpman (1991). Quality ladders in the theory of growth, *Review of Economic Studies* 58, 43-61.

Grossman, H. I. and M. Kim (1996). Predation and accumulation, *Journal of Economic Growth* 1, 333-350.

Grossmann, V. (1998). Are status concerns harmful for growth?, *Finanzarchiv* 55, 357-373.

Grossmann, V. (1999a). Lohnungleichheit in Europa, Arbeitslosigkeit in den USA: Wirklich zwei Seiten derselben Medaille?, *Wirtschaft und Gesellschaft* 25, 423-447.

Grossmann, V. (1999b). Socially optimal redistribution and growth: a note on the welfare significance of the representative consumer, University of Regensburg (mimeo).

Grossmann, V. (2000a). Is it rational to internalize the personal norm that one should reciprocate?, University of Regensburg (mimeo).

Grossmann, V. (2000b). Skilled labor reallocation, wage inequality, and unskilled unemployment, *Journal of Institutional and Theoretical Economics* (forthcoming).

Gupta, D. K. (1977). The socio-economic costs of unemployment and income inequality: a cross-national study, Ph.D. Dissertation, University of Pittsburgh.

Hall, R. E. (1988). Intertemporal elasticity of substitution, *Journal of Political Economy* 96, 339-357.

Hammer, M. and J. Champy (1993). Reengineering the cooperations (New York: Harper Business).

Hanson, G. H. and A. Harrison (1995). Trade, technology and wage inequality, NBER, Working Paper No. 5110.

Harbaugh, R. (1996). Falling behind the Joneses: relative consumption and the growth-savings paradox, *Economics Letters* 53, 297-304.

Harrod, R. F. (1939). An essay in dynamic theory, *Economic Journal* 49, 14-33.

Heckman, J. J., L. Lochner and C. Taber (1998). Explaining rising wage inequality: explorations with a dynamic general equilibrium model of labor earnings with heterogeneous agents, *Review of Economic Dynamics* 1, 1-58.

Hibbs, D. A (1973). Mass political violence: a cross-national causal analysis (New York: Wiley).

Hirsch, F. (1976). Social limits to growth (Cambridge, MA: Harvard University Press).

Hof, F. X. (1999). Relative consumption and endogenous labor supply in the Ramsey model: do status-conscious people work too much?, University of Technology Vienna (mimeo).

Holländer, H. (1996). On the validity of utility statements: standard theory vs Duesenberry's, University of Dortmund (mimeo).

Holsey, C. M. and T. E. Borcherding (1997). Why does the government's share of national income grow? An assessment of the recent literature on the U.S. experience, in: D. C. Mueller (ed.), Perspectives on public choice: a handbook (Cambridge: Cambridge University Press), 562-589.

Hoon, H. T. and E. S. Phelps (1997). Growth, wealth and the natural rate: Is Europe's jobs crises a growth crises? *European Economic Review* 41, 549-557.

Itaya, J. D., E. De Meza, G. D. Myles (1997). In praise of inequality: public good provision and income distribution, *Economics Letters* 57, 289-296.

Johnson, G. E. (1997). Changes in earnings inequality: the role of demand shifts, *Journal of Economic Perspectives* 11, 41-54.

Jones, C. I. (1995). R&D-based models of economic growth, *Journal of Political Economy* 103, 759-784.

Jones, C. I. (1999). Growth: with or without scale effects?, *American Economic Review* 89, 139-144.

Jorgenson, D. W. and K. J. Stiroh (1999). Information technology and growth, *American Economic Review Papers and Proceedings* 89, 109-115.

Jovanovic, B. (1998). Vintage capital and inequality, *Review of Economic Dynamics* 1, 497-530.

Juhn, C., K. M. Murphy and B. Pierce (1993). Wage inequality and the rise in returns to skill, *Journal of Political Economy* 101, 410-442.

Kahneman, D., J. Knetsch and R. Thaler (1986). Fairness as a constraint on profit seeking: entitlements in the market, *American Economic Review* 76, 729-741.

Kakwani, N. C. (1980). Income inequality and poverty: methods of estimation and policy applications (Oxford: Oxford University Press).

Kaldor, N. (1956). Alternative theories of distribution, *Review of Economic Studies* 23, 83-100.

Katz, L. F. and K. M. Murphy (1992). Changes in relative wages, 1963-1987: supply and demand factors, *Quarterly Journal of Economics* 107, 35-78.

Katz, L. F., G. W. Loveman and D. G. Blanchflower (1995). A comparison of changes in the structure of wages in four OECD countries, in: Freeman, R. B. and L. F. Katz (eds.), Differences and changes in wage structures (Chicago: University of Chicago Press).

Kenny, C. (1999). Does growth cause happiness or does happiness cause growth? *Kyklos* 52, 3-26.

Keynes, J. M. (1936). The general theory of employment, interest and money (London: Macmillan).

Knell, M. (1999). Social comparisons, inequality and growth, *Journal of Institutional and Theoretical Economics* (forthcoming).

Kneller, R., M. F. Bleaney and N. Gemmell (1999). Fiscal policy and growth: evidence from OECD countries, *Journal of Public Economics* 74, 171-190.

Konrad, K. (1992). Wealth seeking reconsidered, *Journal of Economic Behavior and Organization* 18, 215-227.

Koopmans, T. C. (1965). On the concept of optimal economic growth, in: Study week on the econometric approach of development planning (Amsterdam: North Holland).

Kremer, M. and E. Maskin (1996). Segregation by skills and the rise in inequality, NBER, Working Paper No. 5718.

Kristov, L. P. Lindert and R. McClelland (1992). Pressure groups and redistribution, *Journal of Public Economics* 48, 135-163.

Krueger, A. B. and L. H. Summers (1987). Reflections on the inter-industry wage structure, in: K. Lang and J. S. Leonard (eds.), Unemployment and the structure of labor markets (New York: Basil Blackwell).

Krueger, A. B. and L. H. Summers (1988). Efficiency wages and the inter-industry wage structure, *Econometrica* 56, 259-293.

Krueger, A. B. (1993). How computers have changed the wage structure: evidence from microdata, 1984-1989, *Quarterly Journal of Economics* 108, 33-60.

Krueger, A. B. (1999). Measuring labor's share, *American Economic Review Papers and Proceedings* 89, 45-51.

Krugman, P. (1994). Past and prospective causes of high unemployment, in: Reducing unemployment. Current issues and options, Proceedings of a Symposium in Jackson Hole, Wyoming, Kansas City: Federal Reserve Bank of Kansas, 68-81.

Külp, B. (1994). Verteilung: Theorie und Politik, 3rd edition (Stuttgart: Gustav Fischer Verlag).

Kuznets, S. (1955). Economic growth and income inequality, *American Economic Review* 45, 1-28.

Lee, W. and J. E. Roemer (1998). Income distribution, redistributive politics and economic growth, *Journal of Economic Growth* 3, 217-240.

Levine, D. I. (1993). Fairness, markets, and ability to pay: evidence from compensation executives, *American Economic Review* 83, 1241-1259.

Lindbeck, A. and D. J. Snower (1996). Reorganization of firms and labor-market inequality, *American Economic Review Papers and Proceedings* 87, 315-321.

Lindbeck, A. and D. J. Snower (1999). Multi-task learning and the reorganization of work, Institute for the study of labor (IZA), Discussion Paper No. 39.

Lloyd-Ellis, H. (1999). Endogenous technological change and wage inequality, *American Economic Review* 89, 47-77.

Lucas, R. E. (1988). On the mechanics of economic development, *Journal of Monetary Economics* 22, 3-42.

Lucas, R. E. (1990). Supply-side economics: an analytical review, *Oxford Economic Papers* 42, 293-316.

Lybeck, J. A. (1986). The growth of government in developed economies (Aldershot: Gower Press).

Machin, S. and A. Manning (1997). Minimum wages and economic outcomes in Europe, *European Economic Review* 41, 733-742.

Machin, S., A. Ryan and J. van Reenen (1996). Technology and changes in skill structure: evidence from an international panel of industries, CEPR, Discussion Paper No. 1434.

Machin, S. and J. van Reenen (1998). Technology and changes in skill structure: evidence from seven OECD countries, *Quarterly Journal of Economics* 113, 1215-1244.

Mandel, E. (1987). Marx, Karl Heinrich, in: J. Eatwell, M. Milgate and P. Newman (eds.), The new Palgrave: a dictionary of economics (London: Macmillan).

Marshall, A. (1895). Principles of economics, 3rd edition (London: Macmillan).

Marx, K. (1890). Das Kapital I - Kritik der politischen Oekonomie: Der Produktionsprozess des Kapitals, 4th edition, edited by F. Engels, reprinted 1969 (Frankfurt/M: Ullstein).

Matsuyama K. (1998). Endogenous inequality, Northwestern University (mimeo).

McDermott, J. (1997). Exploitation and growth, *Journal of Economic Growth* 2, 251-278.

Meltzer, A. H. and S. F. Richard (1981). A rational theory of the size of government, *Journal of Political Economy* 89, 914-927.

Meltzer, A. H. and S. F. Richard (1983). Tests of a rational theory of the size of government, *Public Choice* 41, 403-418.

Michel, L. and J. Bernstein (1994). The state of working America 1994-1995 (Armonk: M. E. Sharpe).

Mincer, J. (1974): Schooling, experience, and earnings (New York: Columbia University Press).

Miniami, R. (1998). Economic development and income distribution in Japan: an assessment of the Kuznets hypothesis, *Cambridge Journal of Economics* 22, 39-58.

Mirlees, J. A. (1971). An exploration in the theory of optimum income taxation, *Review of Economic Studies* 38, 175-208.

Morand, Oliver F. (1999). Endogenous fertility, income distribution, and growth, *Journal of Economic Growth* 4, 331-349.

Mueller, D. C. (1989). Public choice II: a revised edition of public choice (Cambridge: Cambridge University Press).

Mueller, D. C. and P. Murrell (1986). Interest groups and the size of governments, *Public Choice* 48, 125-145.

Murphy, K. M. and R. Topel (1997). Unemployment and nonempoyment, *American Economic Review Papers and Proceedings* 89, 295-300.

Musgrave, R. A. and P. B. Musgrave (1976). Public finance in theory and practice, 2nd edition (New York: McGraw-Hill).

Neven, D. J. and C. A. Wyplosz (1996). Relative prices, trade and restructuring in European industry, CEPR, Discussion Paper No. 1451.

Nickell, S. and B. Bell (1995), The collapse in demand for the unskilled and unemployment across the OECD, *Oxford Review of Economic Policy* 11, 40-62.

Nickell, S. and B. Bell (1996), Changes in the distribution of wages and unemployment in OECD countries, *American Economic Review Papers and Proceedings* 86, 302-308.

Nickell, S. and B. Bell (1997). Would cutting payroll taxes on the unskilled have a significant impact on unemployment?, in: D. J. Snower and G. de la Dehesa (eds.), Unemployment policy: government options for the labour market (Cambridge: Cambridge University Press), 296-328.

OECD (1993). Employment outlook, Paris

OECD (1996). Employment outlook, Paris.

OECD (1997a). Historical statistics 1995, Paris.

OECD (1997b). Employment outlook, Paris.

OECD (1999a). National accounts volume 1: main aggregates 1960-1997, Paris.

OECD (1999b). Industry productivity, Directorate for Science, Technology and Industry, Paris.

OECD (1999c). Employment outlook, Paris.

Okun, A. (1975). Equality and efficiency. The big tradeoff (Washington D.C.: The Brookings Institution).

Palda, F. (1999). Property rights vs. redistribution: which path to national wealth?, *Public Choice* 101, 129-145.

Pen, J. (1971). Income distribution: facts, theories, policies (New York: Praeger Publishers).

Perotti, R. (1993). Political equilibrium, income distribution, and growth, *Review of Economic Studies* 60, 755-776.

Perotti, R. (1994). Income distribution and investment, *European Economic Review* 38, 827-835.

Perotti, R. (1996). Growth, income distribution, and democracy: what the data say, *Journal of Economic Growth* 1, 149-187.

Persson, T. and G. Tabellini (1992). Growth, distribution, and politics, *European Economic Review* 36, 593-602.

Persson T. and G. Tabellini (1994). Is inequality harmful for growth?, *American Economic Review* 84, 600-621.

Piketty T. (1997). The dynamics of the wealth distribution and the interest rate with credit rationing, *Review of Economic Studies* 64, 173-189.

Pissarides, C. A. (1990). Equilibrium unemployment theory (Oxford: Basil Blackwell).

Postlewaite, A. (1998). The social basis for interdependent preferences, *European Economic Review* 42, 779-800.

Poterba, J. M. (1997). The rate of return to corporate capital and factor shares, NBER, Working Paper No. 6263.

Putterman, L., J. E. Roemer and J. Silvestre (1998). Does egalitarianism have a future?, *Journal of Economic Literature* 36, 861-902.

Ramsey, F. (1928). A mathematical theory of saving, *Economic Journal* 38, 543-559.

Rauscher, M. (1997). Conspicuous consumption, economic growth, and taxation, *Journal of Economics* 66, 35-42.

Rebelo, S. (1991). Long-run policy analysis and long-run growth, *Journal of Political Economy* 99, 500-521.

Ricardo, D. (1821). The principles of political economy and taxation, 3rd edition, reprinted 1965 (London: Everyman's Library).

Roberts, K. W. (1977). Voting over income tax schedules, *Journal of Public Economics* 8, 329-340.

Rodrik, D. (1998). Why do more open economies have bigger governments?, *Journal of Political Economy* 106, 997-1032.

Roemer, J. E. (1998). Why the poor do not expropriate the rich: an old argument in new garb, *Journal of Public Economics* 70, 399-424.

Romer P. M. (1986). Increasing returns and long-run growth, *Journal of Political Economy* 94, 1002-1037.

Romer, P. M. (1990). Endogenous technological change, *Journal of Political Economy* 98, S71-S102.

Romer, T. (1975). Individual welfare, majority voting, and the properties of a linear income tax, *Journal of Public Economics* 4, 163-185.

Saint-Paul, G. (1996). Are the unemployed unemployable?, *European Economic Review* 40, 1501-1519.

Saint-Paul, G. and T. Verdier (1992). Education, democracy and growth, *Journal of Development Economics* 42, 399-407.

Saith, A. (1983). Development and distribution: a critique of the cross-country U-hypothesis, *Journal of Development Economics* 13, 367-382.

Sala-i-Martin, X. X. (1996). A positive theory of social security, *Journal of Economic Growth* 1, 277-304.

Sandler, T. (1997). Global challenges: an approach to environmental, political and economic problems (Cambridge: Cambridge University Press).

Sappington, D. (1983). Limited liability contracts between principal and agent, *Journal of Economic Theory* 29, 1-21.

Schlicht, E. (1978). Labour turnover, wage structure, and natural unemployment, *Journal of Institutional and Theoretical Economics* 134, 337-346.

Schlicht, E. (1992). Wage generosity, *Journal of Institutional and Theoretical Economics* 148, 437-451.

Schumpeter, J. A. (1965). Geschichte der ökonomischen Analyse, Band I und II (Göttingen: Vandenhoeck & Ruprecht).

Segerstrom, P. (1998). Endogenous growth without scale effects, *American Economic Review* 88, 1290-1310.

Shackleton, J. R. (1998). Industrial relations reform in Britain since 1979, *Journal of Labor Research* 19, 581-605.

Siebert, H. (1997). Labor market rigidities: a root of unemployment in Europe, *Journal of Economic Perspectives* 11, 37-54.

Slichter, S. (1950). Notes on the structure of wages, *Review of Economics and Statistics* 32, 80-91.

Smith, A. (1776). An inquiry into the nature and causes of the wealth of nations, reprinted 1937 (New York: Random House).

Snower, D. J. (1999a). Causes of changing earnings inequality, Institute for the Study of Labor, Discussion Paper No. 29.

Snower, D. J. (1999b). The organizational revolution and its labor market implications, Keynote address to the European Association of Labour Economists (EALE), Regensburg.

Solow, R. M. (1956). A contribution to the theory of economic growth, *Quarterly Journal of Economics* 70, 65-94.

Solow, R. M. (1979). Another possible source of wage stickiness, *Journal of Macroeconomics* 1, 79-82.

Steindl, J. (1952). Maturity and stagnation in American capitalism (Oxford: Basil Blackwell).

Stigler, G. J. (1941). Production and distribution theories (New York: MacMillan Company).

Stiglitz, J. E. and A. Weiss (1981). Credit rationing in markets with imperfect information, *American Economic Review* 71, 393-410.

Summers, R. and A. Heston (1991). The Penn World Table (Mark 5): An expanded set of international comparisons, 1950-1988, *Quarterly Journal of Economics* 106, 327-368.

Swan, T. W. (1956). Economic growth and capital accumulation, *Economic Record* 32, 334-361.

Thurow, L. C. (1998). Wage dispersion: "Who done it", *Journal of Post Keynesian Economics* 21, 5-37.

Topel, R. H. (1997). Factor proportions and relative wages: The supply-side determinants of wage inequality, *Journal of Economic Perspectives* 11, 55-74.

Uhlig, H. and N. Yanagawa (1996) Increasing the capital tax may lead to faster growth, *European Economic Review* 40, 1521-1540.

Veblen, T. (1899). The theory of the leisure class: an economic study of institutions, reprinted 1934 (New York: Modern Library).

Veenhoven, R. (1995). World database of happiness, *Social Indicators Research* 34, 299-313.

Veenhoven, R. (1998). World database of happiness, Bibliography of happiness, *www.eur.nl.fws/research/happiness.*

Vogt, W. (1999). Skill-biased technological change and inequality: a dynamic interdependence, European Association of Labour Economists (EALE) conference, Regensburg.

Walther, H. (1996). Zurück zur Vollbeschäftigung - aber wie? *Wirtschaft und Gesellschaft* 22, 481-513.

Weiss, Y and C. Fershtman (1998). Social status and economic performance: a survey, *European Economic Review* 42, 801-820.

Welch, F. (1999). In defense of inequality, *American Economic Review Papers and Proceedings* 89, 1-17.

Wicksteed, P. H. (1894). An essay of the co-ordination of the laws of distribution (London: MacMillan & Co.).

WIFO (1999). Employment in European manufacturing, Brussels and Vienna.

World Bank (1998). World Bank economic indicators CD-ROM, Washington D.C.

Young, A. (1993). Invention and bounded learning by doing, *Journal of Political Economy* 101, 443-472.

Young, A. (1998). Growth without scale effects, *Journal of Political Economy* 106, 41-63.

Zweimüller, J. (1998). Schumpeterian entrepreneurs meet Engel's law: the impact of inequality on innovation-driven growth, CEPR, Discussion Paper No. 1880.

Contributions to Economics

Simon Duindam
Military Conscription
1999. ISBN 3-7908-1203-X

Bruno Jeitziner
Political Economy of the Swiss National Bank
1999. ISBN 3-7908-1209-9

Irene Ring et al. (Eds.)
Regional Sustainability
1999. ISBN 3-7908-1233-1

Katharina Müller/Andreas Ryll/Hans-Jürgen Wagener (Eds.)
Transformation of Social Security: Pensions in Central-Eastern Europe
1999. ISBN 3-7908-1210-2

Stefan Traub
Framing Effects in Taxation
1999. ISBN 3-7908-1240-4

Pablo Coto-Millán
Utility and Production
1999. ISBN 3-7908-1153-X

Frank Riedel
Imperfect Information and Investor Heterogeneity in the Bond Market
2000. ISBN 3-7908-1247-1

Kirsten Ralf
Business Cycles
2000. ISBN 3-7908-1245-5

Michele Bagella/Leonardo Becchetti (Eds.)
The Competitive Advantage of Industrial Districts
2000. ISBN 3-7908-1254-4

Frank Bohn
Monetary Union and Fiscal Stability
2000. ISBN 3-7908-1266-8

Michael Malakellis
Integrated Macro-Micro-Modelling Under Rational Expectations
2000. ISBN 3-7908-1266-8

Stefan Baumgärtner
Ambivalent Joint Production and the Natural Environment
2000. ISBN 3-7908-1290-0

Henri Capron, Wim Meeusen (Eds.)
The National Innovation System of Belgium
2000. ISBN 3-7908-1308-7

Tobias Miarka
Financial Intermediation and Deregulation
2000. ISBN 3-7908-1307-9

Chisato Yoshida
Illegal Immigration and Economic Welfare
2000. ISBN 3-7908-1315-X

Nikolaus Thumm
Intellectual Property Rights
2001. ISBN 3-7908-1329-X

Johannes Bröcker/Hayo Herrmann (Eds.)
Spatial Change and Interregional Flows in the Integrating Europe
2001. ISBN 3-7908-1352-4

Kirstin Hubrich
Cointegrated Analysis in a German Monetary System
2001. ISBN 3-7908-1352-4

Nico Heerink et al. (Eds.)
Economic Policy and Sustainable Land Use
2001. ISBN 3-7908-1351-6

Friedel Bolle/Michael Carlberg
Advances in Behavioral Economics
2001. ISBN 3-7908-1358-3

GPSR Compliance

The European Union's (EU) General Product Safety Regulation (GPSR) is a set of rules that requires consumer products to be safe and our obligations to ensure this.

If you have any concerns about our products, you can contact us on ProductSafety@springernature.com

In case Publisher is established outside the EU, the EU authorized representative is:

Springer Nature Customer Service Center GmbH
Europaplatz 3
69115 Heidelberg, Germany

Batch number: 10725973

Printed by Printforce, the Netherlands